INFORMATION SOCIETY

Information has come to be regarded as a symbol of the age in which we live. Talk nowadays is frequently of an 'information explosion', an information technology 'revolution' even of an 'information society'. But just what is an information society and how are we to make sense of it?

This book sets out to examine and assess the variety of theories of information in society currently available. Frank Webster sceptically examines what thinkers mean by an information society, and looks closely at different approaches to informational developments. He provides critical commentaries on the major post-war theories: Daniel Bell's ideas on a post-industrial information society, Anthony Giddens' thoughts on the growth of surveillance and the expansion of the nation state, Herbert Schiller's insistence that information both expresses and consolidates the interests of corporate capitalism: Jürgen Habermas' account of the diminishment of the public sphere; Jean Baudrillard's thoughts on postmodernism and information, and Manuel Castells' depiction of the 'informational city'. Each theorisation is subjected to close scrutiny and is tested against empirical evidence to assess its worth.

The author concludes that, while there has undoubtedly been an information explosion, it is premature to conceive of an information society. We should rather emphasise the 'informatisation' of established relations.

Frank Webster is Professor of Sociology, Oxford Brookes University.

THEORIES OF THE INFORMATION SOCIETY

Frank Webster

London and New York

IN MEMORY OF FRANK NEVILLE WEBSTER
June 20, 1920 – July 15, 1993

First published 1995
by Routledge
11 New Fetter Lane, London EC4P 4EE

Simultaneously published in the USA and Canada
by Routledge
29 West 35th Street, New York, NY 10001

Reprinted in 1996, 1997

Typeset in Times by
Michael Mepham, Frome, Somerset

Printed and bound in Great Britain by
Mackays of Chatham PLC, Chatham, Kent

British Library Cataloguing in Publication Data
A catalogue record for this book is available from the British Library

Library of Congress Cataloguing in Publication Data
Webster, Frank.
Theories of the information society / Frank Webster.
p. cm. – (The international library of sociology)
Includes bibliographical references and index.
ISBN 0-415-10573-0 (hbk)
ISBN 0-415-10574-9 (pbk)
1. Communication—Social aspects.
2. Communication—Technological innovations.
3. Information society. 4. Information technology.
5. Information services and state. I. Title. II. Series.
HM258.W39   1995
302.2—dc20   94-49029
                                       CIP
ISBN 0–415–10573–0 (hbk)
ISBN 0–415–10574–9 (pbk)

CONTENTS

ACKNOWLEDGEMENTS

It has taken me several years to complete this book. Almost all writing is hard, but I have found this exceptionally demanding for a variety of reasons, not all of them academic.

Finishing the book allows me to thank those who have helped me get on with living as well as with writing during some difficult times. Above all that is Liz Chapman and our children Frankie and Isabelle.

My long-time co-author Kevin Robins contributed intellectually and emotionally in inestimable ways.

I'd like to acknowledge too the help of colleagues who supported me when I was slighted in Oxford, especially David Chalcraft whose energy and enthusiasm are infectious and invigorating.

My father was always a central prop of my life. I would have loved to have given him this book as a gift and I am certain he would have been delighted to have received it. Sadly, the dedication is all that I can now offer.

Headington, Oxford, Autumn 1994

1

INTRODUCTION

It is commonplace to describe societies with terms such as 'capitalism', 'feudalism', 'industrialism', 'socialism', 'authoritarianism' and 'totalitarianism'. Most of us will have found ourselves using these sort of words when we have tried to account for events and upheavals, or even for the general drift of change.

In all probability we will have argued with others about the appropriateness of these labels when applied to particular circumstances. For instance, while many will consider Russia in 1995 to have moved well away from the former command economy of Communism, there will be much less agreement that the transition can be accurately described as a shift to a fully capitalist society. There is a constant need to qualify the generalising terminology: hence pre-industrial, advanced capitalism, authoritarian populism, state socialism. . . .

And yet, despite these necessary refinements, few of us will feel able to refuse these concepts or indeed others like them. The obvious reason is that, big and crude and subject to amendment though such *isms* are, they do give us a means to identify and begin to understand essential elements of the world in which we live and from which we have emerged. It seems inescapable that, impelled to make sense of the most consequential features of different societies and circumstances, we are driven towards the adoption of grand concepts.

The starting point for this book is the emergence in recent years of an apparently new way of conceiving contemporary societies. Commentators have increasingly begun to talk about 'information' as a defining feature of the modern world. Much attention is now devoted to the 'informatisation' of social life: we are told that we are entering an 'information age', that a new 'mode of information' predominates, that we have moved into a 'global information economy'. A good many writers have gone even further to identify as 'information societies' the United States, Britain, Japan, Germany and other nations with a similar way of life. Indeed, it appears that information has 'become so important today as to merit treatment as a symbol for the very age in which we live' (Martin, 1988: 303).

Just what sense to make of this symbol has been the source of a very great deal of controversy. To some it constitutes the beginning of a truly professionalised and caring society while to others it represents a tightening of control over the citizenry;

1

to some it heralds the emergence of a highly educated public which has ready access to knowledge while to others it means a deluge of trivia, sensationalism and misleading propaganda; to some it was the development of the nation state which promoted the role of information, while to others it was changes in corporate organisation that had led information to become more critical.

Nonetheless, amidst this extraordinary divergence of opinion, what is striking is that all scholars acknowledge that there is something special about 'information' in the modern era. In an extensive and burgeoning literature concerned with the 'information age' there is little agreement about its major characteristics and its significance other than that – minimally – 'information' has achieved a special pertinence in the contemporary world.

It was curiosity about the currency of 'information' that sparked the idea for this book. It seemed that, on many sides, people were marshalling yet another grandiose term to identify the germane features of our time. But simultaneously thinkers were remarkably divergent in their interpretations of what form this information took, why it was apparently axial to our present systems, and how it was affecting social, economic and political relationships.

It is my intention to focus attention on these different interpretations since it will allow us to scrutinise a common area of interest, even though, as we shall see, interpretations of the role and import of information diverge widely, and, indeed, the closer that we come to examine their terms of reference the less agreement even about the ostensibly common subject matter – information – there appears to be.

Setting out to examine various theories of the 'information society' this book is organised in such a way as to scrutinise major contributions towards our understanding of information in the modern world. For this reason, following an analysis of definitional issues, each chapter looks at a particular theory and its most prominent proponents and attempts to assess its strengths and weaknesses in light of alternative theoretical analyses and empirical evidence.

As they progress through this book readers will encounter Daniel Bell's conception of post-industrial society which places a special emphasis on information (Chapter 3), Anthony Giddens' thoughts on the nation state and violence which highlight the part played by information gathered for surveillance purposes (Chapter 4), Herbert Schiller's views on advanced capitalism's need for and manipulation of information (Chapter 5), Jürgen Habermas' argument that the 'public sphere' is in decline and with it the integrity of information (Chapter 6), the contention that we are living through a transition from Fordist to post-Fordist society that generates and relies upon information handling to succeed (Chapter 7), Jean Baudrillard and other devotees of postmodernism who give particular attention to the explosion of signs in the modern era (Chapter 8), and Manuel Castells' views on the 'informational city' (Chapter 9).

It will not escape notice that these thinkers and the theories with which they are associated, ranging across disciplines such as sociology, philosophy, economics and geography, are at the centre of contemporary debates in social science. This is, of course, not especially surprising given that social thinkers are engaged in trying

2

to understand and explain the world in which we live and that a manifestly important feature of this is change in the informational realm. It is surely unconscionable that anyone should attempt to account for the state of the world without paying due attention to that enormous domain which covers changes in mass media, the spread of information technologies, developments in telecommunications, new forms of work, and even shifts in educational curricula.

However, because this book starts from contemporary social science, it is worth warning that some readers may find at least parts of what follows rather difficult. My experience of teaching the issues, chiefly to advanced level science and technology undergraduates, tells me that it can be disconcerting for students interested in the 'information age' to encounter what to them can appear rather alien and arcane social theorists. They *know* for certain that there has been a radical, even a revolutionary, breakthrough in the technological realm and they want, accordingly, a straightforward account of the social and economic consequences of this development. 'Theory', especially 'grand theory' which has ambitions to identify the most salient features of contemporary life and which frequently has recourse to history and an array of other 'theorists', many of them long dead, does not, and should not, enter into the matter, since all it does is confuse and obfuscate.

I intentionally approach an understanding of information via encounters with major social theorists by way of a riposte to a rash of pronouncements on the 'information age'. Far, far too much of this has come from 'practical' men (and a few women) who, impressed by the 'Information Technology Revolution' or alerted to the prospect of 'information superhighways' by some report or other in *Time* magazine or on Channel 4, have felt able to reel off social and economic consequences that are likely to follow, or that will even inevitably follow. In these frames work will be transformed, education upturned, corporate structures revitalised, democracy itself will be reassessed . . . all because of the 'information revolution'.

Such approaches have infected – and continue to infect – a vast amount of opinion on the 'information society': in bestselling paperback books with catchy titles such as *The Wired Society*, in academic courses designed to consider the 'social effects of the computer revolution', in countless political and business addresses, and in a scarcely calculable amount of journalism that alerts audiences to prepare for upheaval in all aspects of their lives as a result of the coming 'information age'.

An aim of approaching information from an alternative starting point, that of contemporary social theory, is to demonstrate that the *social impact* approaches towards information are hopelessly simplistic and positively misleading for those who want to understand what is going on and what is most likely to transpire in the future. Another aim, perhaps not surprising for a professional social scientist, is to show that social theory, combined with empirical evidence, is an enormously richer, and hence ultimately more practical and useful, way of understanding and explaining recent trends in the information domain.

3

It is as well to say that, while most of the thinkers I examine in this book address informational trends directly, not all of them do so. Thus while Daniel Bell and Herbert Schiller, in their very different ways and with commendable prescience, have been insisting for over a generation that information and communication issues are at the heart of post-war changes, there are other thinkers whom I consider, such as Jürgen Habermas and Anthony Giddens, who give less direct attention to the informational domain. This is neither because they have nothing to contribute to our understanding of information nor because they do not consider it to be important. Rather it is because their terms of debate are different to my focus on the subject of information. For this reason I have felt free to lead off from discussion of, say, Habermas' notion of the public sphere or from consideration of arguments surrounding an alleged shift from Fordism to post-Fordism, more directly towards my interest in informational issues. Since I am not trying to provide a full exposition of particular social theories but am trying to understand the significance of the information domain in the closing decade of the second millennium with the best tools that are available, this does not seem to me to be illegitimate.

Throughout this book there runs an *interrogative* and *sceptical* view of the 'information society' concept itself. This is why, whenever I use the term, I put it in inverted commas to signal my doubts about its accuracy. These even led to misgivings about entitling the book 'Theories of the information society'. The main problem is that the concept 'information society' carries with it an array of suppositions about what has changed and is changing and how this change is being effected. However, the striking thing is that, outside social science, suspicions about the use of the term 'information society' appear to be few. The word is used unproblematically by a wide section of opinion. Recognition of this encouraged me in my choice of title since it meant at least that people would see instantly, at least in very broad terms, what it was about. Nonetheless, let me emphasise that I hope in what follows to shake at least some of the presumptions of those who subscribe to the notion of the arrival of a novel 'information society'.

In my second chapter I subject the concept 'information society' to close scrutiny and there readers will come across major definitional problems with the term, but right at the outset I would draw attention to a major divide that separates many of the thinkers whom I consider in this book. On the one side are subscribers to the notion of an 'information society', while on the other are those that insist that we have only had the 'informatisation' of established relationships. It will become clear that this is not a mere academic division, since the different terminology reveals how one is best to understand what is happening in the informational realm.

It is important to highlight the division of opinion as regards the variable interpretations we will encounter in what follows. On the one hand there are those who subscribe to the notion that in recent times we have seen emerge 'information societies' which are marked by their *differences* from hitherto existing societies. Not all of these writers are altogether happy with the term 'information society', but in so far as they argue that the present era marks a turning point in social development, they can be described as its endorsers. On the other hand there are scholars who, while happy to concede that information has taken on a special significance in the modern era, insist that the central feature of the present is its *continuities* with the past.

I shall be considering various perspectives on 'information' in the contemporary world, discussing thinkers and theories such as Daniel Bell's 'post-industrialism', Jean-François Lyotard on 'postmodernism', and Jürgen Habermas on the 'public sphere'. We shall see that each has a distinct contribution to make to our understanding of informational developments. Nonetheless, beyond and between these differences is a line that should not be ignored: the separation between those who endorse the idea of an 'information society' and those who regard 'informatisation' as the continuation of pre-established relations. Towards one wing we may position those who proclaim a new sort of society that has emerged from the old. Drawn to this side are theorists of:

post-industrialism (Daniel Bell and a legion of followers);
postmodernism (e.g. Jean Baudrillard, Mark Poster);
flexible specialisation (e.g. Michael Piore and Charles Sabel, Larry Hirschhorn);
the informational mode of development (Manuel Castells).

On the other side are writers who place emphasis on continuities. I would include here theorists of:

neo-Marxism (e.g. Herbert Schiller);
Regulation Theory (e.g. Michel Aglietta, Alain Lipietz);
flexible accumulation (David Harvey);
the nation state and violence (Anthony Giddens);
the public sphere (Jürgen Habermas, Nicholas Garnham).

None of the latter denies that information is of key importance to the modern world, but unlike the former they argue that its form and function are subordinate to long-established principles and practices. As they progress through this book readers will have the chance to decide which approaches they find most persuasive.

2

INFORMATION AND THE IDEA
OF AN INFORMATION SOCIETY

Before we can adequately appreciate different approaches to understanding informational trends and issues nowadays we need to pay attention to the definitions which are brought into play by participants in the debates. In particular I think it is especially helpful to examine at the outset what those – and they are everywhere! – who refer to an 'information society' mean when they evoke this term. The insistence of those who subscribe to this concept, and their confident assertion that our time is one marked by its novelty, cries out for analysis, more urgently perhaps than those scenarios which contend that the status quo remains unchanged. The primary aim of this chapter is to ask: what do people mean when they refer to an 'information society'? What criteria do they offer to distinguish an 'information society' from other types? Of course, it is also unavoidable that in examining 'information society' theorists I shall consider aspects of those who refuse the term, since much of any critique of 'information society' notions requires expression of critics' misgivings.

Later I shall move beyond scrutiny of 'information society' definitions to comment on the different ways in which contributors perceive 'information' itself. As we shall see – here, in the very conception of the element which underlies all discussion – there are distinctions which echo the divide between 'information society' theorists who announce the novelty of the present and 'informatisation' thinkers who recognise the force of the past weighing heavily on today's developments.

DEFINITIONS OF THE 'INFORMATION SOCIETY'

It is possible to distinguish *analytically* five definitions of an 'information society' each of which presents criteria for identifying the new. These are:

technological
economic
occupational
spatial
cultural

6

Of course these need not be mutually exclusive (see, for example, Dordick and Wang, 1993), though, as we shall see, particular theorists emphasise one or other factors in presenting their particular scenarios. Let us look more closely at these definitions.

Technological

The most common definition of the 'information society' lays emphasis upon spectacular technological innovation. The key idea is that breakthroughs in information processing, storage and transmission have led to the application of information technologies (IT) in virtually all corners of society. The major concern here is the astonishing reductions in the costs of computers, their prodigious increases in power, and their consequent application any and everywhere (Office of Technology Assessment, 1990: 46). Because it is now economical and feasible to put computers in typewriters, in cars, cookers, watches, factory machines, televisions, kids' toys . . . it follows that we are certain to experience social upheaval of such magnitude that we shall enter a new era. Many books, magazine articles and TV presentations have encouraged the development of a distinct genre which offers this viewpoint: the 'mighty micro' will usher in an entirely new 'silicon civilisation'.

Somewhat more sophisticated versions of this technological route to the 'information society' pay attention to the convergence and imbrication of telecommunications and computing. In these instances the argument runs along the following lines: cheap information processing and storage technologies (computers) result in their being extensively distributed; one of the major areas impacted is telecommunications, notably switching centres which, in being computerised, in effect merge with the general development of computing and impel still more dramatic improvement of information management and distribution. This unification is especially fortuitous because the widespread dissemination of computers means that for optimum use they require connection. In short, the computerisation of telecommunications means that it is increasingly the case that computer can be linked to computer: hence the prospect of links between terminals within and between offices, banks, homes, shops, factories and schools.

This scenario of networked computers is often compared to the provision of electricity: the 'information grid' is seen as analogous to the electrical supply. As the electricity grid links every home, office, factory and shop to provide energy, so the information grid offers information wherever it is needed. This is, of course, an evolutionary process, but with the spread of an ISDN (Integrated Services Digital Network) we have the foundational elements of an 'information society'.

Once established, these information network routes become the highways of the modern age, akin to the roads, railways and canals of the industrial era. As the latter were crucial because they carried back and forth the materials and goods which made the Industrial Revolution, so an ISDN will provide the infrastructure supporting the 'information society'.

7

Undoubtedly what we have here is a technological definition of an 'information society'. Whether it is one which envisages it being born as a result of the *impact* of dramatically new technological innovations or as the outcome of a more incremental development of ISDN systems, all perceive technology as the major distinguishing feature of the new order.

It is very tempting to dismiss technological approaches to the 'information society'. There has been a surfeit of gee-whiz writing that, awed by the pace and magnitude of technological change, naïvely tells us that 'the Computer Revolution. . . . will have an overwhelming and comprehensive impact, affecting every human being on earth in every aspect of his or her life' (Evans, 1979: 13). The genre of futurism which adopts this tone is characteristically full of dire 'wake up' warnings, shallow analyses of the substantive realm, and the self-assurance that only the author has understood what most others have yet to comprehend. It presents a poor case for the validity of technological measures (Webster and Robins, 1986, ch.2).

Nevertheless, if writers such as Alvin Toffler and James Martin impel one towards ready rejection of technological criteria, it has to be acknowledged that very many more serious scholars adopt what is at root a similar approach. In Britain, for example, a much respected school of thought has devised a neo-Schumpeterian approach to change. Combining Schumpeter's argument that major technological innovations bring about 'creative destruction' with Kondratieff's theme of 'long waves' of economic development, these researchers contend that IT represents the establishment of a new epoch. This new 'techno-economic paradigm' (Freeman and Perez, 1988) constitutes the 'Information Age' which is set to mature early in the next century (Hall and Preston, 1988; Freeman, 1987; Freeman *et al.*, 1982).

Elsewhere Michael Piore and Charles Sabel (1984) have suggested that it is the new technologies which provide the foundation for a radically different way of working – 'flexible specialisation'. Thanks to communication and computer technologies, and the information edge they give to small firms now able to quickly assess markets and adroitly respond to them, the prospect is for an end to 'mass production' and its replacement with customised products made by multi-skilled and adaptable craftspeople.

It has to be conceded that, commonsensically, these technological definitions of the 'information society' do seem appropriate. After all, if it is possible to see a 'series of inventions' (Landes, 1969) as the key characteristic of the 'industrial society', then why not accept the virtuoso developments in IT as evidence of a new type of society? As John Naisbitt puts it: 'Computer technology is to the information age what mechanization was to the industrial revolution' (1984: 28). And why not?

Technological definitions of the 'information society' must encounter at least two well-founded objections.

When one reads the literature which tells of profound and portentous changes that new technology is bringing and will bring about one cannot but be struck by the palpable presence of this technology. Whether it is learning about the impact of the 'microelectronics revolution' on the home, CNC (Computer Numerical Control) effects on production processes, or the tidal force of a 'third wave' of computers, telecommunications and biotechnology that announces 'the death knell of industrialism and the rise of a new civilisation' (Toffler, 1980: 2), there is a self-evident reality about the *hereness* of the new technologies. Since this is so, and each of us can see it with our own eyes, then it does seem obvious that the technologies are appropriately regarded as distinguishing features of a new society.

But when one probes further one cannot but be struck also by the astonishing vagueness of technology in most of these books. In *this* society *now* how much IT is there and how far does this take us towards qualifying for 'information society' status? How much IT is required in order to identify an 'information society'? Asking simply for a usable measure, one quickly becomes aware that a good many of those who emphasise technology are not able to provide us with anything so mundanely real-worldly or testable. IT, it begins to appear, is everywhere . . . and nowhere too.

This problem of measurement, and the associated difficulty of stipulating the point on the technological scale at which a society is judged to have entered an 'information age', is surely central to any acceptable definition of a distinctively new type of society. It is ignored by popular futurists: the new technologies are announced and it is unproblematically presumed that this announcement *in and of itself* heralds the 'information society'. This issue is, surprisingly, also bypassed by other scholars who yet assert that IT is the major index of an 'information society'. They are content to describe in general terms technological innovations, somehow presuming that this is enough to distinguish the new society.

There are, however, scholars who are not content with this and find that the issue of measurement causes considerable obstacles to progress. They encounter two particularly awkward problems. First, how does one measure the rate of techno-logical diffusion? and second, when does a society cease being 'industrial' and enter into the 'information' category? These are formidably difficult questions, and ones which should make enthusiasts for the 'information society' scenario hesitate. For instance, in Britain over a decade of social science research by PICT (Programme on Information and Communication Technologies), which was charged with map-ping and measuring the 'information society', has not as yet produced any definitive ways of meeting its objectives (Miles *et al.*, 1990). Certainly there have been some advances, with several studies charting the diffusion of some IT into factories and offices (e.g. Northcott and Walling, 1989). But how is one to assess this diffusion in more general terms: by expenditure on IT (yet given the tumbling prices of the new technologies how is one to differentiate the economic variable from the more central element of information handling capacity?) or by the amount and range of IT introduced? Ought one to centre on IT expenditure or take-up per head or is it better to examine this on an institutional basis? How is one to quantify the significance of the expansion of microcomputer applications *vis-à-vis* mainframe systems? And, if one opts to focus on the uptake of IT, just what is to count as a relevant technology? For instance, should video equipment come before personal computers, networked systems before robotic applications? Further, while one may be able to imagine a time at which some measures of 'informatisation' will have been developed that gain widespread assent, one will still be left with the serious query: where along that graph is the break point which separates the 'information society' from the merely 'advanced industrial'?

The second objection to technological definitions of the 'information society' is very frequently made. Critics object to those who assert that, in a given era, technologies are first invented and then subsequently *impact* on the society, thereby impelling people to respond by adjusting to the new. Technology in these versions is privileged above all else, so it comes to identify an entire social world: the Steam Age, the Age of the Automobile, the Atomic Age, and so no (Dickson, 1974).

The central objection here is not that this is unavoidably technologically determinist – in that technology is regarded as the prime social dynamic – and as such an oversimplification of processes of change. It most certainly is this, but more important is that it relegates into an entirely separate division social, economic and political dimensions of technological innovation. These follow from, and are subordinate to, the premier league of technology which appears to be self-perpetuating though it leaves its impress on all aspects of society.

But it is demonstratively the case that technology is not aloof from the social realm in this way. On the contrary, it is an integral – indeed constitutive – part of the social. For instance, research and development decisions express priorities and from these value judgements particular types of technology are produced. (For example, military projects received substantially more funding than health work for much of the time in the twentieth-century West; not surprisingly a consequence is state-of-the-art weapon systems which dwarf the advances of treatment of the common cold.) Many studies have shown how technologies bear the impress of social values, whether it be in the architectural design of bridges in New York where heights were set that would prevent public transit systems accessing certain areas; the manufacture of cars that testify to the values of private ownership, presumptions about family size (typically two adults, two children), attitudes towards the environment (profligate use of non-renewable energy alongside pollution), status symbols (the Porsche, the Mini, the Rover), and individual rather than public forms of transit; or the construction of houses which are not just places to live, but also expressions of ways of life, prestige and power relations, and preferences for a variety of lifestyles.

There is an extensive literature on this issue, but I do not wish to labour the point *ad nauseam*. All that is required is to state the objection to the *hypostatisation* of technology as applied to the issue of defining the 'information society'. How can it be acceptable to take what is regarded as an asocial phenomenon (technology) and assert that this then defines the social world?

Economic

There is an established sub-division of economics which concerns itself with the 'economics of information'. From within this, and indeed as a founder of this specialism, the late Fritz Machlup (1902–1983) devoted much of his professional life to the goal of assessing the size and growth of the information industries. Machlup's pioneering work, *The Production and Distribution of Knowledge in the United States* (1962), has been seminal in establishing measures of the 'information society' in economic terms.

Machlup attempted to trace the information industries in statistical terms. He distinguished five broad industry groups (broken into fifty sub-branches), namely:

(a) education (e.g. schools, libraries, colleges);
(b) media of communication (e.g. radio and television, advertising);
(c) information machines (e.g. computer equipment, musical instruments);
(d) information services (e.g. law, insurance, medicine);
(e) other information activities (e.g. research and development, non-profit activities).

Working with these sorts of category, it is possible to ascribe an economic value to each and to trace its contribution to gross national produce (GNP). If the trend is for these to account for an increased proportion of GNP, then one may claim to chart the emergence through time of an 'information economy'. This is just what Machlup (1962) proposed in this early study which calculated that 29 per cent of the United States' GNP in 1958 came from the knowledge industries – at that time a remarkable rate of expansion.

As early as the 1960s management guru Peter Drucker was contending that knowledge 1 'has become the foundation of the modern economy' as we have shifted '[f]rom an economy of goods [to]. . . . a knowledge economy' (Drucker, 1969: 249, 247). Today it is commonplace to argue that we have evolved into a society where the 'distinguishing characteristic. . . . is that knowledge and organisation are the prime creators of wealth' (Karunaratne,1986: 52).

Probably the best known – and certainly the most cited – study of the emergence of an 'information economy' conceived on these lines comes in a nine-volume report from Marc Porat (1977b). In allocating industries to his five categories Fritz Machlup had adopted catholic definitions of 'knowledge production', broadly including those which created new information and those which communicated it. Porat echoed much of Machlup's approach in his reliance on government statistical sources to design a computer model of the US economy in the late 1960s, but divided the economy between the 'primary', 'secondary' and 'non-information' sectors. This tripartite schema stemmed from his identification of a weakness in Machlup's, which failed to account for information activities that were disguised from initial examination, for example because they are an in-house element of other industries. Porat included in the 'primary information sector' all those industries which make available their information in established markets or elsewhere where an economic value can be readily ascribed (e.g. mass media, education, advertising, computer manufacture).

However, Porat then sought to identify a 'secondary information sector' which would allow him to include in his typology important informational activities such

as research and development inside a pharmaceutical company, information produced by government departments for internal consumption, and the library
resources of an oil corporation. In this way Porat is able to distinguish the two
information sectors, then to consolidate them, separate out the non-informational
elements of the economy, and, by reaggregating national economic statistics, is able
to conclude that over 46 per cent of the US gross national product is accounted for
by the information sector. *Ipso facto*, 'The United States is now an information-
based economy.' As such it is an 'information society [where] the major arenas of
economic activity are the information goods and service producers, and the public
and private [secondary information sector] bureaucracies' (Porat, 1978: 11).

This quantification of the economic significance of information is an impressive
achievement. It is not surprising that those convinced of the emergence of an
'information society' have routinely turned to Machlup and especially Porat as
authoritative demonstrations of a rising curve of information activity, one set to
lead the way to a new age.

But, there are also difficulties with the economics of information approach
(Monk, 1989: 39–63). One is that, behind the weighty statistical tables that are
resonant of objective demonstration, there is a great deal of hidden interpretation
and value judgement as to how to construct categories and what to include and
exclude from the information sector.

For instance, Marc Porat is at some pains to identify the 'quasi-firm' embedded
within a non-informational enterprise. But is it acceptable, from the correct assumption that R&D in a petrochemical company involves informational activity, to
separate this from the manufacturing element for statistical purposes? It is surely
likely that the activities are blurred, with the R&D section intimately tied to
production wings, and any separation for mathematical reasons is unfaithful to its
role. More generally, when Porat examines his 'secondary information sector' he
in fact splits every industry into the informational and non-informational domains.
But such divisions between the 'thinking' and the 'doing' are extraordinarily hard
to accept – where does one put operation of computer numerical control systems
or the line management functions which are an integral element of production? The
objection here is that Porat divides, arbitrarily, within industries to chart the
'secondary information sector' as opposed to the 'noninformational' realm.

Such objections may not invalidate the findings of Machlup and Porat, and they
are not intended to do that, but they are a reminder of the unavoidable intrusion of
value judgements in the construction of their statistical tables. As such they support
a healthy scepticism about the idea of an emergent 'information economy'.

A second difficulty is that the aggregated data inevitably homogenise very
disparate economic activities. In the round it may be possible to say that growth in
the economic worth of advertising and television is indicative of an 'information
society', but one is left with an urge to distinguish between informational activities
on qualitative grounds. In asking which economically assessed characteristics are
more central to the emergence of an 'information society', one is requesting
scholars to distinguish between, say, information stemming from policy research

12

centres, corporate think tanks, transnational finance houses, manufacturers of 35mm cameras, software designers, and the copywriters of Saatchi and Saatchi.

The enthusiasm of the information economists to put a price tag on everything has the unfortunate consequence of failing to let us know the really valuable dimensions of the information sector. This search to differentiate between *quantitative* and *qualitative* indices of an 'information society' is not pursued by Machlup and Porat, though on a commonsensical level it is obvious that the 4 million sales of the *Sun* cannot be equated with – still less be regarded as more informational than, though doubtless the sales are of more economic value – the 200,000 circulation of the *Financial Times*. It is a distinction to which we shall return, but one which suggests the possibility that we could have a society in which, as measured by GNP, informational activity is of great weight, but of little consequence in terms of the springs of economic, social and political life.

Of course these economists are concerned solely with developing quantitative measurements of the information sector, so the issue of the qualitative worth of information would be of limited relevance to them. However, even on their own terms there are problems. One, mentioned earlier, is the question: 'at which point on the economic graph does one enter an information society?' When 50 per cent of GNP is dedicated to informational activities? This may seem to be a reasonable point, one at which, in straightforward quantitative terms, information begins to predominate. Sadly for 'information society' theorists, however, we are some distance even from that point. Replication studies of Machlup and Porat lead one to qualify any initial sighting of the new age. Michael Rogers Rubin and Mary Taylor Huber (1986) in a large-scale update of Machlup's study concluded that in the United States the contribution of 'knowledge industries' to GNP increased from 28.6 per cent to 34.3 per cent between 1958 and 1980, with virtually no change since 1970, this constituting an 'extremely modest rate of growth relative to the average rate of growth of other components of total GNP' (1986: 3). Furthermore, the same authors' replication of Porat's influential study found little expansion of the information sector during the 1970s when compared to other contributors to GNP. These econometric studies scarcely trumpet the arrival of an 'information society'.

Occupational

A popular measure of the emergence of an 'information society' is the one which focuses on occupational change. Put simply, the contention is that we have achieved an 'information society' when the predominance of occupations is found in information work. That is, the 'information society' has arrived when clerks, teachers, lawyers and entertainers outnumber coalminers, steelworkers, dockers and builders.

The occupational definition is frequently combined with an economic measure. Marc Porat, for example, calculated that by the late 1960s a little under half the US labour force was found in the 'information sector', a growth of almost 500 per cent during a century in which agricultural employment plummeted and information occupations expanded massively. On the surface the changing distribution of jobs seems an appropriate measure. After all, it appears obvious that as work which demands physical strength and manual dexterity such as hewing coal and farming the land declines to be replaced by more and more manipulation of figures and text such as in education and large bureaucracies, we are entering a new type of society. Today 'only a shrinking minority of the labour force toils in factories . . . and the labour market is now dominated by information operatives who make their living by virtue of the fact that they possess the information needed to get things done' (Stonier, 1983: 7, 8).

This trend is seized upon by many reports. For instance, two influential OECD (Organisation for Economic Co-operation and Development) publications (1981, 1986) produced figures from all member countries signalling 'continued growth . . . in those occupations primarily concerned with the creation and handling of information and with its infrastructure support' (1986). Elsewhere Marc Porat identifies an 'astonishing growth rate' of the 'information work force' which doubled every 18.7 years between 1860 and 1980 (Porat, 1977a: 131), thereby propelling the USA towards 'the edge of an information economy' (p.204).

The shift in the distribution of occupations is at the heart of the most influential theory of the 'information society': that of Daniel Bell. Bell sees in the emergence of 'white collar society' (and hence information work) and the decline of industrial labour changes as profound as the end of class-based political conflict, more communal consciousness, and the development of equality between the sexes.

We will consider and critique Bell's theorisation in the next chapter, but here it is appropriate to raise some general objections to occupational measures of the 'information society'. A major problem concerns the methodology for allocating workers to particular categories. The end product – a bald statistical figure giving a precise percentage of 'information workers' – hides the complex processes by which researchers construct their categories and allocate people to one or another.

Marc Porat, for instance, develops what has become an influential typology to locate occupations that are primarily engaged in the production, processing or distribution of information. His is a threefold scheme which encompasses over 400 occupational types that are reported by the US Census and Bureau of Labour Statistics. He explains it as follows:

> The first category includes those workers whose output as primary activity is producing and selling knowledge. Included here are scientists, inventors, teachers, librarians, journalists, and authors. The second major class of workers covers those who gather and disseminate information. These workers move information within firms and within markets; they search, co-ordinate, plan and process market information. Included here are managers, secretaries, clerks, lawyers, brokers, and typists. The last class includes workers who operate the information machines and technologies that support

14

the previous two activities. Included here are computer operators, telephone installers, and television repairers.

(Porat, 1978: 5–6)

Jonscher (1983) simplifies this further still, discerning just two sectors of the economy: the first, an 'information sector', is where people whose prime function is creating, processing and handling information; the second, a 'production sector', is where workers are found who chiefly create, process and handle physical goods.

These distinctions appear reasonable, precise and empirically valid, but there are difficulties. Not least is that 'stating precisely who is an information worker and who is not is a risky proposition' (Porat, 1978: 5). Indeed it is, since every occupation involves a significant degree of information processing and cognition. Porat acknowledges this in his attempt to distinguish non-informational from informational labour on the basis of estimating the *degree* to which each type is involved with information. In other words the categorisation is a matter of judging the extent to which jobs are informational or not. Crude percentages of 'information workers' disguise the fact that they are the outcome of the researcher's estimations (Porat, 1977a: 3).

For example, the railway signalman must have a stock of knowledge about tracks and timetables, about roles and routines; he needs to communicate with other signalmen down the line, with station personnel and engine drivers, is required to 'know the block' of his own and other cabins, must keep a precise and comprehensive ledger of all traffic that moves through his area, and has little need of physical strength to pull levers since the advent of modern equipment. Yet the railway signalman is, doubtless, a manual worker of the 'industrial age'. Conversely, the person who comes to repair the photocopier may know little about products other than the one for which she has been trained, may well have to work in hot, dirty and uncomfortable circumstances, and may need considerable strength to move heavy machinery and replace damaged parts. Yet she will undoubtedly be classified as an 'information worker' since her work with new age machinery suits Dr Porat's interpretations.

The point to be made here is simple: we need to be sceptical of conclusive figures that are the outcomes of researchers' perceptions of where occupations are to be most appropriately categorised. As a matter of fact social scientists know very little about the detail and complexity of people's jobs; there are precious few ethnographies which record the stuff of working lives (see Terkel, 1977). And researchers trying to label 'information' and 'non-information' work are just as much in the dark as the rest of their social science colleagues.

If one may be suspicious of the ways in which researchers conceive information work, one needs also to beware the oversimplifications which can come from allocating a wide variety of jobs to the same pigeonholes. Ian Miles rightly observes in a commentary on these methods that 'the categories of work subsumed under the different headings are often extremely heterogeneous' (Miles, 1991: 917). When one considers, for example, that Marc Porat's first category (information producers) lumps together opticians, library assistants, composers, paperback writers, university professors and engineers, while his second (information distributors) subsumes journalists on quality newspapers with deliverers on the street, and when the OECD puts together as information producers physicists, commodity brokers and auctioneers, then one may well have doubts about the value of this composition of occupations as a means of identifying social change. Further, what of the diversity of occupations each with the same title? Librarian, for example, can encompass someone spending much of the day issuing books for loan and reshelving, as well as someone routinely involved in advising academics on the best sources of information for progressing state-of-the-art research. Is it really sensible to lump together such diversity?

Finally, an important consequence of this homogenisation is a failure to identify the more *strategically* central information occupations. While the methodology may provide us with a picture of greater amounts of information work taking place, it does not offer any means of differentiating the most important dimensions of information work. The pursuit of a quantitative measure of information work disguises the possibility that the growth of certain types of information occupation may have particular consequences for social life.

It has to be said that counting the number of 'information workers' in a society tells us nothing about the hierarchies – and associated variations in power and esteem – of these people. For example, it could be argued that the crucial issue has been the growth of computing and telecommunications engineers, since they may exercise a decisive influence over the pace of technological innovation. A similar, perhaps even greater, rate of expansion in social workers to handle problems of an ageing population, increased family dislocation and juvenile delinquency may have little or nothing to do with an 'information society', though undoubtedly social workers would be classified with IT engineers as 'information workers'.

Again, it could plausibly be argued that the critical factor in the emergence of an 'information society' is the central importance of theoretical, as opposed to practical, knowledge. If one argues that in contemporary society important developments, whether in economics or in technological innovation, commence with established theories (Keynesian, monetarist, etc. in economics; scientific principles in technology), then we must differentiate those who possess this theoretical knowledge from those who are information workers but who merely carry out the practical tasks determined by the theoreticians (such as implementing an economic policy or conducting a laboratory experiment).

Or it may be argued that it is an 'inner circle' (Useem, 1985; Useem and Karabel, 1986) of corporate leaders, quite different from their predecessors, which is the most decisive index of the 'information society'. These are people who are empowered by communicative skills, analytical abilities, foresight and capacities to formulate strategic policies, who also enjoy privileged educational backgrounds, connections through shared clubs and boardroom affiliations, plus access to sophisticated information and communications technologies. All of this provides them with extraordinary leverage over social, economic and political affairs at national and even international level. They are information specialists, but radically different from the run-of-the-mill information workers that crude quantitative methodologists would lump them with.

Perhaps we can better understand this need to qualitatively distinguish between groups of 'information workers' by reflecting on a study by social historian Harold Perkin. In *The Rise of Professional Society* (1990) Perkin argues that the history of Britain since 1880 may be written largely as the rise to pre-eminence of 'professionals' who rule by virtue of 'human capital created by education and enhanced by . . . the exclusion of the unqualified' (p.2). Perkin contends that certified expertise has been 'the organising principle of post-war society' (p.406), the expert displacing once-dominant groups (working-class organisations, capitalist entrepreneurs and the landed aristocracy) and their outdated ideals (of co-operation and solidarity, of property and the market, and of the paternal gentleman) with the professional's ethos of service, certification and efficiency.

To be sure, professionals within the private sector argue fiercely with those in the public, but Perkin insists that this is an internecine struggle, one within 'professional society', which decisively excludes the non-expert from serious participation and shares fundamental assumptions (notably the primacy of trained expertise and reward based on merit).

Alvin Gouldner's discussion of the 'new class' provides an interesting complement to Perkin's. Gouldner identifies a new type of employee that has expanded in the twentieth century, a 'new class' that is 'composed of intellectuals and technical intelligensia' (Gouldner, 1979: 153) which, while in part self-seeking and often subordinate to powerful groups, can also contest the control of established business and party leaders. Despite these potential powers, the 'new class' is itself divided in various ways. A key division is between those who are for the most part technocratic and conformist and the humanist intellectuals who are critical and emancipatory in orientation. To a large extent this difference is expressed in the conflicts identified by Harold Perkin between private and public sector professionals. For instance, we may find that accountants in the private sector are conservative while there is a propensity for humanistic intellectuals to be radical.

My main point here is that both Gouldner and Perkin are identifying particular changes within the realm of information work which have especially important consequences for society as a whole. To Gouldner the 'new class' can provide us with vocabularies to discuss and debate the direction of social change, while to Perkin the professionals create new ideals for organising social affairs.

If one is searching for an index of the 'information society' in these thinkers one will be directed to the quality of the contribution of certain groups. Whether one agrees or not with either of these interpretations the challenge to definitions of an 'information society' on the basis of a count of raw numbers of 'information workers' should be clear. To thinkers such as Perkin and Gouldner the quantitative change is *not* the main issue. Indeed, as a proportion of the population the groups they lay emphasis upon, while they have expanded, remain distinct minorities. Tiny in the case of Michael Useem's 'inner circle', and more numerous where the growth of professions is identified, but never more than 20 or 25 per cent of the workforce.

Spatial

This conception of the 'information society', while it draws on sociology and economics, has at its core the geographer's distinctive stress on space. Here the major emphasis is on the information *networks* which connect locations and in consequence have dramatic effects on the organisation of time and space.

John Goddard (1992) identifies four interrelated elements in the transition to an 'information society':

1 Information is coming to occupy centre stage as the 'key strategic resource' on which the organisation of the world economy is dependent.

    The modern world demands the co-ordination of globally distributed manufacture, planning across and between sovereign states, and marketing throughout continents. Information is axial to these diverse activities, thus of heightened importance in the contemporary world. It follows too that 'information management' is of exceptional pertinence and that as a result we witness the rapid expansion of information occupations.

2 Computer and communications technologies provide the infrastructure which enables information to be processed and distributed. These technologies allow information to be handled on an historically unprecedented scale, facilitate instantaneous and 'real-time' trading, and monitoring of economic, social and political affairs on a global stage.

3 There has been an exceptionally rapid growth of the 'tradeable information sector' of the economy, by which Professor Goddard means to highlight the explosive growth of services such as new media (satellite broadcasting, cable, video) and on-line data bases providing information on a host of subjects ranging from stock market dealings, commodity prices, patent listings and currency fluctuations, to scientific and technological journal abstracts.

    Complementing these developments has been the radical reorganisation of the world's financial system which has resulted in the collapse of traditional boundaries that once separated banking, brokerage, financial services, credit agencies and the like. Inside this bewildering world of high finance – which few people understand and still fewer appear able to control – circulate, in electronic form, dazzling sums of capital.

4 The growing 'informatisation' of the economy is facilitating the integration of national and regional economies.

    Courtesy of immediate and effective information processing and exchange, economics has become truly global, and with this has come about a reduction in the constraints of space. Companies can now develop global strategies for production, storage and distribution of goods and services, financial interests operate continuously, respond immediately, and traverse the globe. The boundaries erected by geographical location are being pushed further and further back – and with them too the limitations once imposed by time – thanks to the virtuoso ways in which information can be managed and manipulated in the contemporary period.

Added together these trends emphasise the centrality of *information networks* linking together locations within and between towns, regions, nations, continents and the entire world.

The analogy with the electricity grid to which we referred when considering technological definitions of an 'information society' is often drawn here. As the electricity grid runs throughout an entire nation, extending down to the individual householder's ring main, so too we may envisage a 'wired society' (Martin, 1979) operating at the national, international and global level to provide an 'information ring main' (Barron and Curnow, 1978) to each home, shop or office. Increasingly we are all connected to the network – which itself is expanding its reach and capacities.

Many writers put emphasis on the technological bases of the information network (e.g. Hepworth, 1989). Perhaps predictably then with these accounts of an emerging 'network society' considerable attention is given to advances in and obstacles to the development of an ISDN (Integrated Services Digital Network) infrastructure (Dordick *et al.*, 1981).

However, notwithstanding the importance of technology, and actually providing a salutary reminder of the easily neglected centrality of telecommunications to IT developments, most thinkers concerned with the emergence of a 'network marketplace' place stress on ways in which networks underline the significance of the *flow of information* (Castells, 1989)

The salient idea here is of information circulating along electronic 'highways'. Interestingly, no one has been able to quantify how much and at what rate information must flow along these routes to constitute an 'information society'. In fact, no one has produced reliable figures capable of giving us an overall understanding of information traffic (cf. OECD, 1988). We have data on telephone density in relation to population, figures on the expansion of facsimile services, statistics for sales of computer systems, automated telecommunications exchanges and so on, but lack a clear picture of the size, capacity and use of the networks.

Nevertheless, all observers are aware of a massive increase in transborder data flows, in telecommunications facilities, in communications between computers at every level from home to transnational organisation, in exchanges between stock markets and corporate segments, in access to international data bases, in telex messages. Similarly there is considerable awareness of increases in the global distribution of mass-mediated information, satellite television being the obvious and pre-eminent example, though one would have to include news gathering and distribution services in any adequate picture. As Geoff Mulgan has it, 'the networks carry an *unimaginable* volume of messages, conversations, images and commands' (Mulgan, 1991: 1, my emphasis).

Why much greater volume and velocity of information flows should impel us to think in terms of the constitution of a new type of society returns us to the geographer's special concern with space. All things happen in particular places and at specific times, but the characteristics of space and time have been transformed with the advent of the 'network society'. Where once trade was cumbersome and slow-moving across distances, nowadays it can be effected instantaneously with computerised communications technologies; where once corporate activity had to be co-ordinated by slow-moving letter which took days and even weeks to cross the space that divided the interested parties, nowadays it takes place in real time courtesy of sophisticated telecommunications and video conference facilities.

In short, the constraints of space have been dramatically limited, though certainly not eliminated. And simultaneously time has itself been 'shrunk' as contact via computer communications and telecommunications is immediate. This 'time/space compression', as Anthony Giddens terms it, provides corporations, governments and even individuals with hitherto unachievable options.

No one could deny that information networks are an important feature of contemporary societies: satellites do allow instantaneous communications round the globe, data bases can be accessed from Oxford to Los Angeles, Tokyo and Paris, facsimile machines and interconnected computer systems are a routine part of modern businesses.

Yet we may still ask: why should the presence of networks lead analysts to categorise societies as 'information economies'? And when we ask this we encounter the problem of the imprecision of definitions once again. For instance, when is a network a network? Two people speaking to one another by telephone or computer systems transmitting vast data sets through a packet switching exchange? When an office block is 'wired' or when terminals in the home can communicate with local banks and shops? The question of what actually constitutes a network is a serious one and it raises problems not only of how to distinguish between different levels of networking, but also of how we stipulate a point at which we have entered a 'network/information society'.

It also raises the issue of whether we are using a technological definition of the 'information society' – i.e. are networks being defined as technological systems? – or whether a more appropriate focus would be on the flow of information, which for some writers is what distinguishes the present age. If it is the former, then we could take the spread of ISDN technologies as an index, but few scholars offer any guidance as to how to do this. And if it is the latter, then it may reasonably be asked how much and why more volume and velocity of information flow should mark a new society?

Finally, one could argue that information networks have been around for a very long time. From at least the early days of the postal service, through to telegram and telephone facilities, much economic, social and political life is unthinkable without the establishment of such information networks. Given this long-term dependency and incremental, if accelerated, development, why should it be that in the 1980s commentators begin to talk in terms of 'information societies'?

Cultural

The final conception of an 'information society' is perhaps the most easily acknowledged, yet the least measured. Each of us is aware, from the pattern of our everyday lives, that there has been an extraordinary increase in the information in social circulation. There is simply a great deal *more* of it about than ever before.

Television has been in extensive use for over thirty years in Britain, but now its programming is pretty well round-the-clock, people being able to watch from breakfast time until the early morning. It has expanded from a single channel and discontinuous service to include now four broadcast channels (with a fifth soon to come). And this has been enhanced to incorporate video technologies, cable and satellite channels, and even computerised information services such as teletext. There is very much more radio output available now than even a decade ago, at local, national and international level. And radios are no longer fixed in the front room, but spread through the home, in the car, the office, and, with the Walkman, everywhere. Movies have long been an important part of people's information environment and, indeed, attendances at cinemas have significantly declined. But movies are today very much more prevalent than ever: available still at cinema outlets, broadcast on television, readily borrowed from video rental shops, cheaply purchased from the shelves of chain stores. Walk along any street and it is almost impossible to miss the advertising hoardings, the billboards, the window displays in shops. Visit any railway or bus station and one cannot but be struck by the widespread availability of paperback books and inexpensive magazines – their subject matter ranging from classical through middlebrow and self-therapy to pulp fiction – a scale and scope without precedent. In addition, audiotape, compact disc and radio all offer more, and more readily available, music, poetry, drama, humour and education to the general public. Newspapers are extensively available and a good many new titles fall on our doorsteps as free sheets. Junk mail is delivered daily. . . .

All of this testifies to the fact that we inhabit a media-laden society, but the informational features of our world are more thoroughly penetrative than a short list of television, radio, and other media systems suggests. This sort of listing implies that new media surround us, presenting us with messages to which we may or may not respond. But in truth the informational environment is a great deal more intimate, more constitutive *of* us, than this suggests. Consider, for example, the telephone, which is readily accessible to the vast majority of people nowadays. It is absolutely crucial to the organisation of our everyday lives: fixing up a babysitter, contacting the electrician, making sure grandad is OK. . . . Without such information technologies we could function only with the greatest of difficulty (which is the lot of the minority without such facilities). Reflect too that virtually every home has at least one camera which is used to construct a record of significant events (marriages, birthdays, celebrations . . .). Increased numbers own video cameras which capture these experiences still more palpably. Walk into any home today and one sees on its walls, in albums, on the TV monitor, media representations of those

21

who live there and those who are important but absent. These images are not just pictures of family and friends: they *are* the biographies and identities of these people.

One may also consider the informational dimensions of the clothes we wear, the styling of our hair and faces, the very ways in which nowadays we *work at* our image (from body shape to speech, people are intensely aware of the messages they may be projecting and how they feel about themselves in certain clothes, with a particular hairstyle, etc.). A few moments' reflection on the complexities of fashion, the intricacy of the ways in which we design ourselves for everyday presentation, makes one well aware that social intercourse nowadays involves a greater degree of informational content than previously.

Homes too are informational laden in an historically singular way. Furniture, layout, and decorative design all express ideas and ideals: the G-plan style, the Laura Ashley sofa, the William Morris wallpaper . . . and the mixing of some and all of these according to choice and budget. Certainly, since the days of the Industrial Revolution, homes have signified ways of life – one thinks, for example, of the style of the 'respectable' working class of the late Victorian period or the distinctive design of the professional middle classes between the wars. But it is the explosion in variety in recent decades, and the accessibility of it to a great many, that is most remarkable. With this has come an astonishing vista of signification.

This intrusion of information into the most intimate realms of home, bedroom and body is complemented by the growth of institutions dedicated to investing everyday life with symbolic significance. One thinks of the global advertising business, of publishing empires, of the fashion industry, of world-wide agencies of media production which bring to the domestic scene reflections of our own ways of life and images of other lifestyles, thereby presenting us with alternative meanings which may be absorbed, rejected and reinterpreted by people, but all the time adding to the vocabulary of the symbolic environment.

Readers will recognise and acknowledge this extraordinary expansion of the informational content of modern life. Contemporary culture is manifestly more heavily information laden than any of its predecessors. We exist in a media-saturated environment, which means that life is quintessentially about symbolisation, about exchanging and receiving – or trying to exchange and resisting reception – messages about ourselves and others. It is in acknowledgement of this explosion of *signification* that many writers conceive of our having entered an 'information society'. They rarely attempt to gauge this development in quantitative terms, but rather start from the 'obviousness' of our living in a sea of signs, one fuller than at any earlier epoch.

Paradoxically, it is perhaps this very explosion of information which leads some writers to announce, as it were, the death of the sign. Blitzed by signs all around us, designing ourselves with signs, unable to escape signs wherever we may go, the result is, oddly, a collapse of meaning. As Jean Baudrillard puts it: 'there is more and more information, and less and less meaning' (1983a: 95). In this view signs once had a reference (clothes, for example, signified a given status, the political statement a distinct philosophy, the TV news 'what really happened'). However, in this, the 'postmodern' era, we are enmeshed in such a bewildering web of signs that they lose their salience. Signs come from so many directions, and are so diverse, fast-changing and contradictory, that their power to signify is dimmed. In addition, audiences are creative, self-aware and reflective, so much so that all signs are greeted with scepticism and a quizzical eye, hence easily inverted, reinterpreted and refracted from their intended meaning.

As people's knowledge through direct experience declines it becomes evident that signs are no longer straightforwardly representative of something or someone. The notion that signs represent some 'reality' apart from themselves loses its credibility. Rather signs are self-referential: they – *simulations* – are all there is. They are, again to use Baudrillard's terminology, the 'hyper-reality'.

People appreciate this situation readily enough: they deride the poseur who is dressing for effect, but acknowledge that it's all artifice anyway; they are sceptical of the politicians who 'manage' the media and their image through adroit PR, but accept that the whole affair is a matter of information management and manipulation. Here it is conceded that people do not hunger for any true signs, because they recognise that there are no longer any truths. In these terms we have entered an age of 'spectacle' in which people realise the artificiality of signs they may be sent ('it's only John Major at his latest photo opportunity', 'it's news manufacture', 'it's Jack playing the tough guy') and in which they also acknowledge the inauthenticity of the signs they use to construct themselves ('I'll just put on my face'; 'there I was adopting the ''worried parent'' role').

As a result signs lose their meaning and people simply take what they like from those they encounter (usually very different meanings than may have been intended at the outset). And then, in putting together signs for their homes, work and selves, they happily revel in their artificiality, 'playfully' mixing different images to present no distinct meaning, but instead to derive 'pleasure' in the parody or pastiche of, say, combining punk and a 1950s Marilyn Monroe facial style. In this 'information society' we have then 'a set of meanings [which] is communicated [but which] have no meaning' (Poster, 1990: 63).

Experientially this idea of an 'information society' is easily enough recognised, but as a definition of a new society it is considerably more wayward than any of the notions we have considered. Given the absence of criteria we might use to measure the growth of signification in recent years it is difficult to see how students of postmodernism such as Mark Poster (1990) can depict the present as characterised by a novel 'mode of information'. How can we know this other than from our sense that there is more symbolic interplay going on? And on what basis can we distinguish this society from say, that of the 1920s, other than purely as a matter of degree of difference? As we shall see (in Chapter 8) those who reflect on the 'postmodern condition' have interesting things to say about the character of contemporary culture, but as regards establishing a clear definition of the 'information society' they are glaringly deficient.

QUALITY AND QUANTITY

Reviewing these varying definitions of the 'information society' what becomes abundantly clear is that they are either or both underdeveloped or imprecise. Whether it is a technological, economic, occupational, spatial or cultural conception, we are left with highly problematical notions of what constitutes, and how to distinguish, an 'information society'.

It is important that we remain aware of these difficulties. Though as a heuristic device the term 'information society' has some value in exploring features of the contemporary world it is far too inexact to be acceptable as a definitive term. For this reason, throughout this book, though I shall on occasion use the concept and will consistently acknowledge that information plays a critical role in the present age, I will express suspicion as regards 'information society' scenarios and remain sceptical of the view that information has become the major distinguishing feature of our times.

For the moment, however, I want to raise some further difficulties with the language of the 'information society'. The first problem concerns the quantitative versus qualitative measures to which I have already alluded. My earlier concern was chiefly that quantitative approaches failed to distinguish more strategically significant information activity from that which was routine and low-level and that this homogenisation was misleading. Here I again want to raise the quality/quantity issue in so far as it bears upon the question of whether the 'information society' marks a *break* with previous sorts of society.

Most definitions of the 'information society' offer a quantitative measure (numbers of white-collar workers, percentage of GNP devoted to information, etc.) and assume that, at some unspecified point, we enter an 'information society' when this begins to predominate. But there are no clear grounds for designating as a new type of society one in which all we witness is greater quantities of information in circulation and storage. If there is just more information, then it is hard to understand why anyone should suggest that we have before us something radically new. This is a point made well by Anthony Giddens when he observes that all societies, as soon as they are formed into nation states, are 'information societies' in so far as routine gathering, storage and control of information about population and resources are essential to their operation (Giddens, 1985: 178). On this axis all that differentiates the present era from, say seventeenth-century England, is much greater quantities of information that are amassed, dissembled and processed.

Against this, however, it may be feasible to describe as a new sort of society one in which it is possible to locate information of a qualitatively different order and function. Moreover, this does not even require that we discover that a majority of the workforce is engaged in information occupations or that the economy generates a specified sum from informational activity. For example it is theoretically possible to imagine an 'information society' where only a small minority of 'information experts' hold decisive power. Kurt Vonnegut created this image years ago in his novel *Player Piano* and it is the stuff of an entire sub-genre of science fiction. One need look only to the writings of H. G. Wells to conceive of a society in which a 'knowledge elite' predominates and the majority, surplus to economic requirement, are condemned to drone-like unemployment. On a quantitative measure, say of occupational patterns, this would not qualify for 'information society' status, but we could feel impelled to so designate it because of the decisive role of information/knowledge to the power structure and direction of social change.

The blunt point is that quantitative measures – simply more information – cannot of themselves identify a break with previous systems, while it is at least theoretically possible to regard small but decisive qualitative changes as marking a system break. Yet what is especially odd is that so many of those who identify an 'information society' as a new type of society do so by presuming that this qualitative change can be defined simply by calculating how much information is in circulation, how many people work in information jobs and so on. What we have here is the assumption that quantitative increases transform – in unspecified ways – into qualitative changes in the social system.

It is noticeable that those scholars such as Herbert Schiller and David Harvey who stress the present's continuities with the past, while they acknowledge the increasingly central role played by information, have at the forefront of their minds the need to differentiate between categories of information and the purposes to which it is put. In other words, those who insist that the 'informationalised' society is *not* radically different from the past are at pains to differentiate information on qualitative grounds. For instance, they will examine how information availability has been affected by the application of market criteria and contend that the wealthier sectors of society gain access to particularly high-quality information which consolidates their privileges and powers. Yet while they emphasise this sort of qualitative dimension of informatisation they do so to highlight continuities of the socio-economic system. Conversely, those who consider that the 'information society' is a radically different system most often have recourse to quantitative indices to demonstrate a profound qualitative change.

Theodore Roszak (1986) provides an interesting insight into this paradox in his critique of 'information society' themes. His examination emphasises the importance of qualitatively distinguishing 'information', extending to it what each of us does on an everyday basis when we differentiate between phenomena such as data, knowledge, experience and wisdom. Certainly these are themselves slippery terms, but they are an essential part of our daily lives. In Roszak's view the present 'cult of information' functions to destroy these sort of qualitative distinctions which are the stuff of real life. It does this by insisting that information is a purely quantitative thing, subject to statistical measurement. But to achieve calculations of the economic value of the information industries, of the proportion of GNP expended on information activities, the percentage of national income going to the information professions and so on, the qualitative dimensions of the subject (is the information useful? is it true or false?) are laid aside. 'For the information theorist, it does not matter whether we are transmitting a fact, a judgement, a shallow cliché, a deep teaching, a sublime truth, or a nasty obscenity' (Roszak, 1986: 14). These qualitative issues are laid aside as information is homogenised and made amenable to numbering: 'Information comes to be a purely quantitative measure of communicative exchanges' (ibid.: 11).

The astonishing thing to Roszak is that along with this quantitative measure of information comes the assertion that more information is profoundly transforming social life. Having produced awesome statistics on information activity by blurring the sort of qualitative distinctions we all make in our daily lives, 'information society' theorists then assert that these trends are set to change qualitatively our entire lives.

Roszak vigorously contests these ways of thinking about information. A result of a diet of statistic upon statistic about the uptake of computers, the data-processing capacities of new technologies, and the creation of digitalised networks is that people readily come to believe that information is the essential sustenance of the social system. There is so much of this food that it is tempting to agree with those 'information society' theorists who insist that we have entered an entirely new sort of system. But against this 'more-quantity-of-information-to-new-quality-of-society' argument Theodore Roszak insists that the 'master ideas' (p.91) which underpin our civilisation are not based upon information at all. Principles such as 'all men are created equal', 'my country right or wrong', 'live and let live', 'we are all God's children', and 'do unto others as you would be done by' are central ideas of our society – but all come *before* information.

It is important to say that Roszak is not arguing that these and other 'master ideas' are necessarily correct (in fact a good many are noxious – e.g. 'all Jews are rich', 'all women are submissive', 'blacks have natural athletic ability'). But what he is emphasising is that ideas, and the necessarily qualitative engagement these entail, take precedence over quantitative approaches to information. And what he especially objects to is that 'information society' theorists reverse that situation at the same time as they smuggle in the (false) idea that more information is fundamentally transforming the society in which we live.

WHAT IS INFORMATION?

Roszak's rejection of statistical measures leads us to consider perhaps the most significant feature of approaches to the 'information society'. We are led here largely because his advocacy is to reintroduce qualitative judgement into discussions of information. Roszak asks questions like: is more information necessarily making us a better informed citizenry? does the availability of more information make us better informed? what sort of information is being generated and stored and what value is this to the wider society? what sort of information occupations are expanding, why and to what ends?

What is being proposed here is that we insist on examination of the *meaning* of information. And this is surely a commonsensical understanding of the term. After all, the first definition of information that springs to mind is the *semantic* one:

26

information is meaningful; it has a subject; it is intelligence or instruction about something or someone.

If one were to apply this concept of information to an attempt at defining an 'information society' it would follow that we would be discussing these characteristics of the information. We would be saying that information about *these* sorts of issues, *those* areas, *that* economic process, are what constitutes the new age. However, it is precisely this commonsensical definition of information which the 'information society' theorists jettison. What is in fact abandoned is the notion that information has a semantic content.

The definitions of the 'information society' we have reviewed perceive information in non-meaningful ways. That is, searching for quantitative evidence of the growth of information, a wide range of thinkers have conceived it in the classic terms of Claude Shannon and Warren Weaver's (1964) information theory. Here a distinctive definition is used, one which is sharply distinguished from the semantic concept in common parlance. In this theory information is a quantity which is measured in 'bits' and defined in terms of the probabilities of occurrence of symbols. It is a definition derived from and useful to the communications engineer whose interest is with the storage and transmission of symbols, the minimum index of which is on/off (yes/no or 0/1).

This approach allows the otherwise vexatious concept of information to be mathematically tractable, but at the price of excluding the equally vexing – yet crucial – issue of meaning and, integral to meaning, the question of the information's quality. On an everyday level when we receive or exchange information the prime concerns are its meaning and value: is it significant, accurate, absurd, interesting, adequate or helpful? But in terms of the information theory which underpins so many measures of the explosion of information these dimensions are irrelevant. Here information is defined independent of its content, seen as a physical element as much as is energy or matter. As one of the foremost 'information society' devotees puts it:

> *Information exists.* It does not need to be *perceived* to exist. It does not need to be *understood* to exist. It requires no intelligence to interpret it. It does not have to have *meaning* to exist. It exists.
>
> (Stonier, 1990: 21)

In fact, in these terms, two messages, one which is heavily loaded with meaning and the other which is pure nonsense, can be equivalent. As Roszak says, here '*information* has come to denote whatever can be coded for transmission through a channel that connects a source with a receiver, regardless of semantic content' (1986: 13). This allows us to quantify information, but at the cost of abandonment of its meaning and quality.

If this definition of information is the one which pertains in technological and spatial approaches to the 'information society' (where the quantities stored, processed and transmitted are indicative of the sort of indexes produced) we come across a similar elision of meaning from economists' definitions. Here it may not be in terms of 'bits', but at the same time the semantic qualities are evacuated and replaced by the common denominator of price (cf. Arrow, 1979).

To the information engineer the prime concern is with the number of yes/no symbols; to the information economist it is with their vendability. But as the economist moves from consideration of the concept of information to its measurement what is lost is the heterogeneity that springs from its manifold meanings. The 'endeavour to put dollar tags on such things as education, research, and art' (Machlup, 1980: 23) unavoidably abandons the semantic qualities of information. Kenneth Boulding observed a generation ago that 'The bit . . . abstracts completely from the content of information. . . . and while it is enormously useful for telephone engineers . . . for purposes of the social system theorist we need a measure which takes account of significance and which would weight, for instance, the gossip of a teenager rather low and the communications over the hot line between Moscow and Washington rather high' (Boulding, 1971). How odd then that economists have responded to the qualitative problem which is the essence of information with a quantitative approach that, reliant on cost and price, is at best 'a kind of qualitative guesswork' (ibid.). 'Valuing the invaluable', to adopt Fritz Machlup's terminology, means substituting information content with the measuring rod of money. We are then able to produce impressive statistics, but in the process we have lost the notion that information is *about* something (Maasoumi, 1987).

Finally, though culture is quintessentially about meanings, about how and why people live as they do, it is striking that with the celebration of the non-referential character of symbols by enthusiasts of postmodernism we have a congruence with communications theory and the economic approach to information. Here too we have a fascination with the profusion of information, an expansion so prodigious that it has lost its hold semantically. Symbols are now *everywhere* and generated *all* of the time, so much so that their meanings have 'imploded', hence ceasing to signify.

What is most noteworthy is that 'information society' theorists, having jettisoned meaning from their concept of information in order to produce quantitative measures of its growth, then conclude that such is its increased economic worth, the scale of its generation, or simply the amount of symbols swirling around, that society must encounter profoundly meaningful change. We have, in other words, the assessment of information in non-social terms – it just *is* – but we must adjust to its social consequences. This is a familiar situation to sociologists who often come across assertions that phenomena are aloof from society in their development (notably technology and science) but that they carry within them momentous social consequences. It is demonstrably inadequate as an analysis of social change (cf. Dickson, 1974; Woolgar, 1985).

Doubtless being able to quantify the spread of information in general terms has some uses, but it is certainly not sufficient to convince us that in consequence of an expansion society has *profoundly* changed. For any genuine appreciation of what an 'information society' is like, and how different – or similar – it is to other social systems we must surely examine the meaning and quality of the information. What sort of information has increased? Who has generated what kind of information, for what purposes and with what consequences?

As we shall see, scholars who start with these sort of questions, sticking resolutely to questions of the meaning and quality of information, are markedly different in their interpretations from those who operate with non-semantic and quantitative measures. The former are extremely sceptical of alleged transitions to a new age. Certainly they accept that there is more information today, but because they refuse to see this outside of its content (they always ask: *what information?*) they are reluctant to agree that its generation has brought about the transition to an 'information society'.

CONCLUSION

This chapter has raised serious doubts about the validity of the notion of an 'information society'. On the one hand, we have encountered a variety of criteria which purport to measure the emergence of the 'information society'. We shall see later that there are thinkers who, using different criteria, still argue that we have or are set to enter an 'information society'. One can scarcely have confidence in a concept when its defenders diagnose it in quite different ways. Moreover, these criteria, ranging from technology to occupational changes to spatial features, though they appear at first glance robust, are in truth vague and imprecise, incapable on their own of establishing whether or not an 'information society' has arrived or will arrive some time in the future.

On the other hand – and this must make one deeply sceptical of the 'information society' scenario (while not for a moment doubting that there has been an extensive 'informatisation' of life) – is the recurrent shift of its proponents from seeking quantitative measures of the spread of information to the assertion that these indicate a qualitative change in social organisation. The same procedure is evident too in the very definitions of information that are in play, with 'information society' subscribers endorsing non-semantic definitions. These – so many 'bits', so much economic worth – are readily quantifiable, thereby alleviating analysts of the need to raise qualitative questions of meaning and value. But as they do so they fly in the face of commonsensical definitions of the word, conceiving information as devoid of content. Again as we shall see, those scholars who commence their accounts of transformations in the informational realm in this way are radically different from those who, while acknowledging an explosion in information, insist that we never abandon questions of its meaning and purpose.

THE INFORMATION SOCIETY AS POST-INDUSTRIALISM: DANIEL BELL

Amongst those who subscribe to the notion that a new sort of society is emerging, far and away the best known characterisation of the 'information society' is Daniel Bell's theory of post-industrialism. Indeed, the terms are very frequently used interchangeably: the information age is presented as expressive of post-industrial society (hereafter PIS) and post-industrialism is widely regarded as an 'information society'. And, it should be said, Professor Bell, though he coined the term post-industrialism as long ago as the late 1950s, did himself take to substituting the words 'information' and 'knowledge' for the prefix 'post-industrial' around 1980 when a tidal wave of enthusiasm for futurology was swelled by interest in developments in computer and communications technologies.

To be fair, Professor Bell (born 1919), for long an eminent American sociologist (see Liebowitz, 1985), had from the outset of his interest in PIS underlined the central role of information/knowledge for his emergent social system.[1] *The Coming of Post-Industrial Society* seemed to fit quite beautifully with the explosive technological changes that hit advanced societies in the late 1970s. Impacted by the sudden arrival, apparently from out of the blue, of staggering microelectronic technologies which rapidly permeated offices, industrial processes, schools and the home – computers soon seemed to be everywhere – there was an understandable and urgent search to discover where all these changes were leading. With, as it were, a ready-made model available in Daniel Bell's weighty *The Coming of Post-Industrial Society* (1973), we should perhaps not be surprised that so many commentators took it straight from the shelf. PIS just seemed to be *right* as a description of the coming world. In its prescience it gave intellectual order to an unsettling period of change. Given the circumstances, few people seemed prepared to heed Bell's qualification that 'the concept of post-industrial society is only on the level of abstraction' (Bell, 1976a: x).

The renowned professor appeared to have foreseen the turmoil that computer communications technologies especially were bringing into being. Indeed, he had written earlier of the need for a massive expansion of these information technologies and here they were apparently fulfilling his prognosis. Understandably, then, he got the credit and was considered something of a guru. In such circumstances, perhaps Bell's opportunism is understandable and we can sympathise with his jumping on the bandwagon when he too began to adopt the fashionable language of the 'information revolution'.

Writing in the 1990s it is not difficult to pick holes in a conception which has been open to scrutiny for twenty-five years and more. Precious little social science lasts even a decade, so Daniel Bell's continuing to set the terms for such an important debate is an enviable achievement. Nonetheless, I shall be arguing that PIS is untenable and that there is solid social science evidence to demonstrate this.

Moreover, it is especially necessary to argue that PIS is an unhelpful way of understanding the role and significance of information in the present because Bell's image of post-industrialism is so often appropriated by shallow commentators on the 'information society'. They seem to say 'this is a 'post-industrial information society' [such is the formulation of former *Sunday Times* editor Andrew Neil, 1983] – for heavyweight elaboration see distinguished Harvard professor Daniel Bell's 500 page tome'. Such an appeal gives authority, insight and gravitas to articles, books and television specials which offer wild and groundless propositions about the direction and character of the present times and which deserve little serious attention. To be able to demonstrate that PIS is an untenable notion is therefore to undermine a plank of more popular commentary on the conditions in which we find ourselves.

Bell contends that we are entering a new system, a post-industrial society, which, while it has several distinguishing features, is characterised throughout by a heightened presence and significance of information. Daniel Bell argues that information and knowledge are crucial for PIS both quantitatively and qualitatively. On the one hand, features of post-industrialism lead to greater amounts of information being in use. On the other hand, Bell claims that in the post-industrial society a qualitative shift is evident, especially in the rise to prominence of what he calls 'theoretical knowledge'. In PIS, in other words, there is not just more information; there is also a different kind of information/knowledge in play. With such features, it will be readily appreciated why Bell's theory of 'post-industrialism' appeals to those who want to explain the emergence of an 'information society'.

In my view, he is undeniably correct in his perception of increases in the part played by information in social, economic and political affairs. However, Daniel Bell is mistaken in interpreting this as signalling a new type of society – a 'post-industrial' age. Indeed, PIS is unsustainable once one examines it in the light of substantive social trends. Further, it is only sustainable as an 'ideal type' construct by adopting a particular theoretical starting point and methodological approach to social analysis which is shown to be faulty when one comes to look at real social relations. In short, the whole project is deeply flawed empirically, theoretically and methodologically, as the remainder of this chapter will demonstrate.

NEO-EVOLUTIONISM

Daniel Bell believes that the United States leads the world on a path towards a new type of system. Though he does not claim outright that the development of PIS is an inevitable outcome of history, he does think it is possible to trace a movement from pre-industrial, through industrial, to post-industrial societies. There is a distinctive trajectory being described here and it obviously holds to a loose chronology. Certainly it is not difficult to apply Bell's terms to historical periods. For example, Britain in the early eighteenth century was pre-industrial – i.e. agricultural; by the late nineteenth century it was distinctively industrial – i.e. manufacturing was the emphasis; and, as we approach the end of the twentieth century, signs of post-industrialism are clear for all to see – i.e. services predomi-nate. It is hard, looking at Bell's route planning, to resist the view that the motor of history is set on automatic, headed unstoppably towards a fully fledged PIS. Indeed, Bell is confident enough of its direction to contend that post-industrialism 'will be a major feature of the twenty-first century in the social structures of the United States, Japan, the Soviet Union, and Western Europe' (Bell, 1976a: x).

Evolutionist thinking has been out of favour for a long time in social science circles. Redolent as it is of Social Darwinism, of that complacent and rather smug attitude that we (authors of books who happen to live comfortably in the richest countries of the world) inhabit a society towards which all other, less fortunate, ones should aspire and are moving anyway, evolutionism can be hard to defend. It is intellectually vulnerable to at least two serious charges. The first is the fallacy of *historicism* (the idea that it is possible to identify the underlying laws or trends of history and thereby to foresee the future). The second is the trap of *teleological* thinking (the notion that societies change towards some ultimate goal). In contem-porary terms, evolutionist thinking – and some critics would say Bell is an evolutionist – suggests that history has identifiable trends of development in the direction of Western Europe, Japan and, especially, the United States. It follows from this that, somehow, people do not have to do anything, or even worry much about, the problems they encounter in their own societies – injustices, inequalities, the fickleness or obduracy of human beings – because the logic of history ensures that they move inexorably onwards and upwards towards a better and more desirable order.

Daniel Bell is too sophisticated a thinker to fall for these charges. He is quick to repudiate such accusations, though denial alone does not ensure innocence.[2] Certainly it is difficult to avoid the conclusion that PIS is a superior form of society to anything that has gone before, just as it is hard to resist the idea that we are moving ineluctably towards 'post-industrialism' due to underlying social trends. When I review below Bell's description of PIS, readers will be able to gauge this commitment to evolutionist premises for themselves.

SEPARATE REALMS

But first an important theoretical and methodological point which is fundamental to Daniel Bell's *oeuvre*. PIS emerges through changes in *social structure* rather than in politics or culture. Its development most certainly 'poses questions' (Bell,

1976a: 13) for the polity and cultural domain, but Bell is emphatic that change cannot be seen to be emanating from any one sector to influence every other dimension of society. In his view advanced societies are 'radically disjunctive' (Bell, 1980: 329). That is, there are independent 'realms' – social structure, polity and culture – which have an autonomy one from another such that an occurrence in one realm cannot be presumed to shape another.

Put in other terms, Bell is an *anti-holist*, iterating over and again that societies are *not* 'organic or so integrated as to be analyzable as a single system' (Bell, 1976a: 114). He rejects all totalistic/holistic theories of society, whether (and especially) they come from the Left and conceive of capitalism as something which intrudes into each and every aspect of society, or whether they are more conservative and believe society functions in an integrative manner, tending towards order and equilibrium. Against these approaches Bell divides, apparently arbitrarily and certainly without explicit reasoning (why just three realms? why not an independent realm for law, family or education?), contemporary societies into the three realms of social structure, politics and culture.

It might be wondered if this point is really worth making. Why bother with Bell's insistence that societies are divided into separable realms? The reason is that it is absolutely pivotal for several aspects of Bell's thought. For instance, this radical separation of realms enables Bell to sidestep awkward questions of the degree to which developments in any one realm exert influence on another. He can concede that there are 'questions' posed by events in one sphere for others – but he goes no further than this, concluding that his concern is only with one particular realm. And that is surely not acceptable. Since Bell can insist that the realms are independent he can evade the awkward issue of stipulating the inter-realm relationships by returning again and again to his theoretical and methodological premiss.

Again, Bell offers us no evidence or argument to justify his starting point (Ross, 1974: 332–334). Since in the everyday world of human existence issues inevitably pose themselves in ways which involve the interconnections of culture, politics and social structure, it is surely at the least evasive for Bell to insist on their 'radical disjuncture'.

Finally, one of the most striking features of Bell's account of PIS is that it reveals the breakdown of a one-time 'common value system' (Bell, 1976a: 12) which held throughout society, but which is now being destroyed. Indeed, he insists that 'in our times there has been an increasing disjunction of the three [realms]' (p.13). In a later work, *The Cultural Contradictions of Capitalism* (1976b), the breakdown of a once complementary cultural ethos and social structural requirements (the nineteenth-century Protestant character structure which conjoined with the demands of socio-economic development) provides Bell with his organising theme. Furthermore, in *The Coming of Post-Industrial Society*, Bell highlights trends, such as the increased presence of professionals, which have important consequences for politics (the common query: *will professionals rule?*). In drawing attention to such issues Bell is surely underlining the significance, *not* of the disjunction of realm, but of their interconnectedness. How did a once unified culture and social structure

33

come apart and, another side of the same coin, how many linkages remain? If developments in one realm really do have consequences for another, then just what is their nature (Steinfels, 1979: 169)?

POST-INDUSTRIAL SOCIETY

PIS emerges from changes in the social structure only. This includes the economy, the occupational structure and the stratification system, but excludes politics and cultural issues. *The Coming of Post-Industrial Society* is therefore an account of changes taking place in one sector of society only – and one must not presume, says Bell, that these are the most consequent parts.

Bell offers a typology of different societies which is dependent on the predominant mode of *employment* at any one stage. In his view the type of work which is most common becomes a defining feature of particular societies. Thus Bell suggests that while in pre-industrial societies agricultural labour is pretty well ubiquitous, and in industrial societies factory work is the norm, in post-industrial societies it is service employment which predominates.

Why these changes should have happened is explained by Bell when he identifies increases in productivity as the key to change. The critical factor in moving from one society to another is that it becomes possible to get 'more for less' from work because of the application of the principle of 'rationalisation' (efficiency). In the pre-industrial epoch everyone had to work the land just to eke out a subsistence existence. However, as it becomes feasible to feed an entire population without everyone working on the land (for example through improved agricultural practices), so it becomes possible to release a proportion of the people from farms so that they may do other things while still being assured of an adequate food supply. Accordingly, they drift to the towns and villages to supply growing factories with labour while buying their food from the excess produced in the country.

With the progression of this process, we eventually enter the industrial era where factory labour begins to predominate. And always the 'more for less' principle tells. Hence industrial society thrives by applying more and more effective techniques in the factories, which in turn leads to sustained increases in productivity. Steam power reduces the need for muscle power while increasing output; electricity allows assembly lines to run.... The history of industrialisation can be written in this way of the impressive march of mechanisation and automation which guaranteed spectacular increases in productivity. The indomitable logic is more output from fewer and fewer workers.

As productivity soars surpluses are produced from the factories, which enables expenditures to be made on things once unthinkable luxuries: for example, teachers, hospitals, entertainment, even holidays. In turn, these expenditures of industrial-earned wealth create employment opportunities in services, occupations aimed at satisfying new needs that have emerged, and have become affordable, courtesy of industrial society's bounty. The more wealth industry manages to create, and the fewer workers it needs to do this thanks to technical innovations (the familiar motor of 'more for less'), the more services that can be afforded and the more people that can be released from industry to find employment in services.

So long as this process continues – and Bell insists that it is ongoing as we enter PIS – then we are assured of:

- a decline of workers employed in industry, ultimately reducing to a situation where very few people find work in industry (the era of 'robotic factories', 'total automation' etc.);
- accompanying this decline in industrial employment, continuing and sustained increases in industrial output because of unrelenting rationalisation;
- continued increases of wealth, translated from industry's output, which may be spent on new needs people may feel disposed to originate and fulfil (anything from hospital facilities to masseurs);
- continuous release of people from employment in industrial occupations;
- creation of a never-ending supply of new job opportunities in services aimed at fulfilling the new needs that more wealth generates (i.e. as people get richer they discover new things to spend their money on and these require service workers).

Bell's identification of post-industrialism draws on familiar empirical social science. It is undeniably the case – as detailed as long ago as 1940 by Colin Clark and quantified later by, amongst others, Victor Fuchs (1968) – that there has been a marked decline of primary (broadly agricultural and extractive industries) and secondary (manufacturing) sector employment and a counterbalancing expansion of tertiary, or service sector, jobs. For Bell, a 'service society' is a post-industrial one too.

But before elaborating that we must emphasise that service sector employment is, in a very real sense, the end of a long history of transfers of employment from one sector to another. The reasoning behind this is straightforward: the ethos of 'more for less' impels automation of, first agriculture, and later on industry, thereby getting rid of the farmhand and later on the industrial working class while simultaneously ensuring increased wealth. To thinkers like Bell these redundancies are a positive development since, towards the end of the 'industrial society' era, they at once gets rid of unpleasant manual labour and, simultaneously abolish radical politics since, asks Bell, how can the class struggle be waged when the proletariat is disappearing? At the same time, while automation abolishes the working class it still leaves the wider society in receipt of continually expanding wealth. And society, receiving these additional resources, develops novel needs which use up these additional resources. Society is richer? New needs are imagined? These result in continually increasing services, for example hotels, tourism, and psychiatry. Indeed, it should be noted that needs are truly insatiable. Provided there is money to spend, people will manage to generate additional needs: masseurs, participative sports, psychotherapists. . . . Moreover, service employment has a distinctive trait which makes it especially difficult to automate. Since it is person-oriented and usually intangible, productivity increases courtesy of machines are not really feasible. How does one begin to automate a social worker, nurse or teacher?

In short, services will increase the more productivity/wealth is squeezed out of agriculture and industry, but there is not much fear that jobs in services will themselves be automated. Because of this, an evolutionary process that has told decisively throughout the pre-industrial and industrial epochs loses its force as we find ourselves in a mature PIS. With the coming of the post-industrial society we reach an end of history as regards job displacement caused by technical innovations.

THE ROLE OF INFORMATION

If one can accept that sustained increases in wealth result in service jobs predominating, one may still wonder where information comes into the equation. Why should Bell feel able to boldly state that 'the post-industrial society is an information society' (1976a: 467) and that a 'service economy' indicates the arrival of post-industrialism? It is not difficult to understand information's place in the theorisation. Bell explains with a number of connected observations. Crucially it involves the character of life in different epochs. In pre-industrial society life is 'a game against nature' where 'one works with raw muscle power' (Bell, 1976a: 126); in the industrial era, where the 'machine predominates' in a 'technical and rationalized' existence, life 'is a game against fabricated nature' (ibid.). In contrast to both, life in a 'post-industrial society [which] is based on services . . . is a game between persons' (p.127). Here 'what counts is not raw muscle power, or energy, but information'.

Where once one had struggled to eke a living from the land and had to rely on brawn and traditional ways of doing things (pre-industrialism), and where later one was tied to the exigencies of machine production (industrialism), with the emergence of a service/post-industrial society the material of work for the majority is information. After all, a 'game between persons' is necessarily one in which information is the basic resource. What do bankers do but handle money transactions? What do therapists do but conduct a dialogue with their clients? What do advertisers do but create and transmit images and symbols? What do teachers do but communicate knowledge? Service work surely *is* information work. Necessarily then, the predominance of service employment leads to greater quantities of information. To restate this in Bell's later terminology, it is possible to distinguish three types of work:'extractive', 'fabrication' and 'information activities' (Bell, 1979,p.178),the balance of which has changed over the centuries so that in PIS the 'predominant group [of occupations] consists of information workers' (p.183).

Daniel Bell, however, goes further than this to depict PIS as an especially appealing place to live, for several reasons. First of all, information work is mostly white-collar employment which, since it involves dealing with people rather than with things, brings promise of greater job satisfaction than hitherto. Second, within the service sector professional jobs flourish, accounting, Bell claims, for more than 30 per cent of the labour force by the late 1980s (Bell, 1989: 168).[3] This means that the 'central person' in PIS 'is the professional, for he is equipped, by his education and training, to provide the kinds of skill which are increasingly demanded in the post-industrial society' (1976a: 127). Third, 'the core of the post-industrial society is its professional technical services' (Bell, 1987: 33), the 'scientists and engineers, who form the key group in the post-industrial society' (Bell, 1976a: 17). Fourth, it is a particular segment of services which 'is decisive for post-industrial society'.This consists of those professionals in health, education, research and government, where we are able to witness 'the expansion of a new intelligentsia – in the universities, research organizations, professions, and government' (ibid.: 15).

More professional work, more role for the intellectuals, more importance placed on qualifications, and more person-to-person employment. Not only does this provide an especially appealing prospect; it also promotes the role of information/knowledge. I shall return to this, but should note here that Bell pushes even further the positive features of PIS. As far as he is concerned, the rise of professionals means not only that a great deal more information is in circulation, but also that society undergoes decisive qualitative changes. One reason for this is that professionals, being knowledge experts, are disposed towards planning. As this disposition becomes a more dominant feature of the society, so it displaces the vicissitudes of *laissez-faire*. Because professionals will not leave the future to the anarchy of the free market, replacing the hidden hand with forecasts, strategies and plans, PIS develops a more intentional and self-conscious developmental trajectory, thereby taking control of its destiny in ways previously unimaginable. A second qualitative change revolves around the fact that, since services are 'games between people' conducted by professionals, the quality of this relationship comes to the forefront. Scholars are not concerned with the profit and loss they stand to make on an individual student; what matters is the development of the young person's knowledge, character and skills. The doctor does not regard the patient as 'x' amount of income. Further, and logically following, this person-oriented society in which professionals' knowledge is so telling evolves into a *caring* society. In 'post-industrial' society people are not to be treated as units (the fate of the industrial worker in an era when concern was with machinery and money), but rather will benefit from the person-oriented services of professionals that are premissed on the needs of the client. The imperative to plan alongside this impulse to care leads, says Bell, to a 'new consciousness' in PIS which, as a 'communal society' (1976a: 220), promotes the 'community rather than the individual' (p.128) as the central reference point. Concerns like the environment, care of the elderly, the achievements of education which must be more than vocational, all take precedence over mere matters of economic output and competitiveness – and, thanks to the professionals' expertise and priorities, can be addressed. They represent a shift, attests Bell, from an 'economizing' (maximisation of return for self-interest) ethos towards a 'sociologizing' mode of life ('the effort to judge a society's needs in more conscious fashion' . . . on the basis of some explicit conception of the "public interest" ' (Bell, 1976a: 283).

Readers may at this point be reminded of the request to reflect on the charge that the theory of 'post-industrial' society contains evolutionary assumptions. It is, I think, hard to avoid the conclusion that PIS is a superior form of society, one at a higher stage of development than its predecessors, and one towards which all societies capable of increasing productivity are moving.

INTELLECTUAL CONSERVATISM

What is clear in all of this is that increases in information work and a greater availability of professional occupations lead Daniel Bell to identify a distinctive *break* between industrial and 'post-industrial' societies. While it is incontestable that there is more information employment than hitherto, and that there is an obvious escalation of information in use, there are major problems with Bell's argument that 'post-industrialism' marks a system break with previous societies.

One difficulty is with the rather shaky foundations on which Bell constructs his theory of a new type of society. There is no inherent reason why increases in professionals, even striking ones, should lead one to conclude that a new age is upon us. For instance, it seems perfectly reasonable to suppose that if, say, the pattern of industrial ownership remained the same and the dynamic which drove the economy stayed constant, then the system – occupations apart – would remain intact. No one has suggested, for example, that a country such as Switzerland, because it is heavily reliant on banking and finance, is a fundamentally different society to, say, Norway or Spain where occupations are differently spread. All are recognisably capitalist, whatever surface features they may exhibit.

Bell and his sympathisers have two responses to this. The first revolves around the question: what degree of change does one need to conclude there has been a systemic break? The only honest answer to this is that it is a matter of judgement and reasoned argument – and I shall produce reasons to support my judgement of systemic continuity in a moment. Second, it must be conceded that Bell, with his commitment to separate 'realm' analysis, could reply that changes along one axis represent a new social order even while on other, unconnected, dimensions there are continuities. *Ipso facto* his commitment to there being an identifiable 'post-industrial' society evidenced by occupational and informational developments could be sustained. I shall reply to these defences below (p.40ff.) by arguing that his anti-holism is untenable and that it is possible to demonstrate that there are identifiable continuities which have a systemic reach.

But before we proceed to these more substantial arguments, there is another reason to suspect the idea of a new 'post-industrial' era emerging. This may be explored by examining the reasons Bell offers by way of explanation of the transition from the old to the new regime. When we ask *why* these changes occur, Bell appeals to arguments which are remarkably familiar in social science research. Such is this intellectual conservatism that we have grounds to be sceptical about the validity of his claim that a radically new system is emerging.

As we saw earlier, the reason for change, according to Bell, is that increases in productivity allow employees to shift from agriculture and industry to services.

Productivity increases come from technological innovations which gave us more food from fewer farmers and more goods from factories with fewer workers. As Bell says, 'technology . . . is the basis of increased productivity, and productivity has been the transforming fact of economic life' (1976a: 191), and it is this productivity which lays the basis for PIS since its beneficence pays for all those service occupations.

What is noticeable about this is that it is a very familiar form of sociological reasoning and, being an expression of *technological determinism*, one which is deeply suspect in social science. It carries two especially dubious implications: that technologies are the decisive agents of social change; and that technologies are themselves aloof from the social world, though they have enormous social effects. Where, critics ask, are people, capital, politics, classes, interests in all of this (Webster and Robins, 1986, ch. 2)? Can it seriously be suggested that technologies are at once the motor of change and simultaneously untouched by social relations?

More important than details of the objection to technological determinism here is the need to fully appreciate the more general character of Bell's intellectual conservatism. Presenting this old proposition (traceable back at least to St Simon writing at the early stages of *industrialisation* in the closing years of the eighteenth century), which is heavily criticised in virtually every sociology primer, leads one to query Bell's assertion of the novelty of 'post-industrialism'.

Another source of his views reinforces this suspicion. This is his indebtedness to Max Weber – a major founder of classical sociology who wrote in the late nineteenth and early twentieth centuries of the *industrial* changes taking place around him – and in particular his interpretation of Weber as the major thinker on 'rationalisation'. Bell tells us that Weber thought 'the master key of Western society was rationalization' (Bell, 1976a: 67), which means the growth of an ethos of 'more for less'. To Bell, increases in productivity, indeed the application of new technologies themselves, is at root all a matter of 'rationalization': 'the axial principle of the social structure is *economizing* – a way of allocating resources according to principles of least cost, substitutability, optimization, maximization, and the like' (p.12). Bell is offering here a remarkably familiar – and vigorously contested – account of change (cf. Janowitz, 1974). It does allow him to deny the charge of technological determinism, but only because there is a dynamic of change still more foundational – rationalisation.

It might be objected that it is possible to be intellectually conservative while still satisfactorily explaining radical social change to a new type of society. However, Bell's dependence on themes central to nineteenth-century social scientists whose concern was to explore the emergence and direction of *industrialism* surely undermines his case that PIS is novel. It is odd, to say the least, to borrow arguments from classical social theorists that were developed to understand the development of industrialism, only to assert that they actually account for the emergence of a new, post-industrial, society (Kumar, 1978: 237).

The emphasis on the role of 'rationalization' leads Bell down a number of well-trodden paths, each of which carries large warning signs from fellow social scientists. Prominent amongst these is that, from his argument that all industrial societies 'are organized around a principle of functional efficiency whose desideratum is to get "more for less" and to choose the more "rational" course of action' (Bell, 1976a: 75–76), he is inevitably endorsing a *convergence theory* of development which ignores, or at least makes subordinate to this 'rationalization', differences of politics, culture and history (Kleinberg, 1973). Insisting that there are 'common characteristics for all industrial societies: the technology is everywhere the same; the same kind of technical and engineering knowledge (and the schooling to provide these) is the same; classification of jobs and skills is roughly the same' (p.75), Bell necessarily contends that all societies are set on the same developmental journey, one which *must* be followed *en route* to PIS.

Another, related, difficulty with this is the problem Bell has in reconciling his view that the productivity gains from the social structure (the 'economizing' mode of industrial societies) must be sustained to enable continued expansion of the service sector, which in turn generates a 'sociologizing' or community consciousness. Since he tells us that the latter will become a defining feature of PIS, and will be accompanied by an outlook sceptical of mere economic output, while simultaneously the economy must expand to support PIS, we are left with a puzzle: are we still mired in 'industrial society', even with multitudes of service workers, where the bottom line is still 'more for less', or have we really transcended this mentality? In answer one must note that we can scarcely be talking about a *post*-industrial society when the continued existence and development of an automated and productive industrial system is a requisite of all the 'post-industrial' changes Professor Bell envisages.

POST-INDUSTRIAL SERVICE SOCIETY?

There seem to me still more reasons for rejecting Bell's depiction of 'post-industrial society'. These can be understood on closer analysis of what Bell takes to be the major sign of the emergence of PIS: the growth of services. I shall demonstrate the *continuities* with established relations which the expansion of services represents, quite in contrast with Bell's postulate that it indicates a *break* with the past. As I do this, by reviewing what may be termed the *Gershuny critique* after its most authoritative formulator, we shall see again that the concept of 'post-industrial society' is unsustainable.

To recapitulate: Professor Bell cites the undeniable fact that the service sector of the economy has expanded while industrial and agricultural sectors have declined as *prima facie* evidence of the coming of 'post-industrialism'. Logically, it seems clear that, with services continuing to grow, and within services professional occupations expanding especially fast,[4] then, provided sufficient wealth can be generated from productivity increases in agriculture and industry by efficiency increases, ultimately almost everyone will find employment in services. This is certainly the conclusion Bell draws from his historical review: he cites figures which show that in 1947 over half the US workforce was in the 'goods-producing' sectors and 49 per cent in the service sector; by 1980 this was projected to change to 32 per cent and 68 per cent respectively (Bell, 1976a: 132). And this trajectory has been verified by the course of events, with every data set subsequently produced demonstrating an expansion of the service sector as a percentage of the total employment (OECD figures show that by 1991 72 per cent of the US workforce was occupied in services: pp.92–93).

It is important that we understand the reasoning being applied here. Bell is dividing employment into three separate sectors – primary, secondary, tertiary (broadly, agriculture, manufacture, services) – but he is also decisively linking them in the following way. He is arguing that services are *dependent* on the outputs from the other two sectors in so far as services consume resources while agriculture and manufacturing generate them. Put in more vulgar terms, he is assuming that the wealth-creating sectors of society must subsidise the wealth-consuming realms. This is, of course, a very familiar nostrum: for example, schools and hospitals must spend only what 'we can afford' from the wealth created by industry.

A key point to be grasped is that Bell is not simply taking the classification of employment into different sectors as indicative of the rise of a 'post-industrial' society. He is also operating with an aetiology, a theory of causation, which underpins the statistical categories. This is frequently unstated, but it is ever-present, and it is the assumption that increased productivity in the primary and secondary sectors is 'the motor that drives the transformative process' (Browning and Singelmann, 1978: 485) towards a service-dominated 'post-industrial' era. Unfortunately for Professor Bell, this presumption is demonstrably false.

The first problem is that Bell's stages view of development – from pre-industrial, to industrial, finally reaching post-industrialism – is historically cavalier. Just as the 'over-tertiarisation' of Third World countries, now regarded as a sign of maladjustment, suggests there is no historical necessity that an industrial base be founded for services, so too is there little evidence to support the notion that advanced societies have progressed from of majority employment in industrial production to majority employment in services. The most spectacular change has not been one of transfer from factory to service employment, but *from agriculture to services*. Moreover, even in Britain, historically the most industrialised of countries, the proportion of the labour force occupied in manufacture was remarkably stable at 45–50 per cent between 1840 and 1980, and it is the collapse of manufacturing industry due to recession and government policies which has dramatically reduced this proportion to less than one-third.

All this is to say that talk of evolutionary shifts from one sector to the next is at the least dubious. Other than in Britain, nowhere has a majority of the population at any time worked in industry, and even in this country it is hard to sustain the argument that employment has shifted in any sequential way. To be sure, the theory of 'post-industrial' society could account for the more common practice of employment transfer from agriculture to services by positing a 'leapfrog' explanation. That is, such is the rapidity of automation that a society may jump from pre-industrialism to post-industrialism in the course of a generation or so because productivity advances in both agriculture and industry are unbounded. In this case, while one may retain doubts about Bell's theme of 'from goods to services', it is possible to hang on to the axial idea that expanded services emanate from the bounty of productivity growth in the other two sectors.

It is the second criticism of Bell's conviction that wealth must be created in agriculture and industry as a prerequisite of service expansion which is most telling. A starting point for this attack is the observation that 'services' is a residual category of statisticians interested in examining employment by economic sectors, something which accounts for anything not classifiable in the primary or secondary sectors and which has been described as 'a rag-bag of industries as different as real-estate and massage parlors, transport and computer bureaux, public administration and public entertainment' (Jones, 1980: 147). The point in stressing the generality and left-over constitution of service industries is that the classificatory convenience which separates the tertiary sector from others is grossly misleading. It is the *social construction* of the category 'services' as industries apart from – yet dependent upon – the fruits of manufacture and agriculture which misleads and allows Bell to suggest, with superficial force, that services will expand on the basis of increased productivity in the primary and secondary sectors. But, it is only at a conceptual level that the service sector can be regarded as distinct from yet dependent on other areas of society.

This becomes clear when, following Jonathan Gershuny, we explore further the meaning of services. Paradoxically, Daniel Bell's theory of 'post-industrial' society replete with services nowhere explicitly defines what a service is. Throughout Bell's writing the service sector is contrasted with the industrial, and we are told that PIS arrives with a switch 'from goods to services', but what a service actually is is not made clear. However,

> it becomes obvious by contrast with the nature of goods: goods are material, permanent, made by people using machines, which are sold or otherwise distributed to people who thereafter may use them at will. Services, we infer by contrast, are immaterial, impermanent, made by people for people.
>
> (Gershuny, 1978: 56)

Bell's entire theory of PIS as a distinctly different stage of development requires that service work is perceived as the *opposite* of goods production, because it is the supply of services (perceived as 'games between people', informational and intangible) which distinguishes PIS from 'industrial' society where most workers were employed in the fabrication of things. It is Bell's thesis that a society moves out of industrialism when it has sufficient wealth to lay out on immaterial services, which in turn generate service occupations that account for the majority of employment and that do not produce goods, but instead consume resources created elsewhere.

The premiss of this model of society and social change is challenged when one examines the substance of service work (i.e. services in terms of occupations – rather than sectoral categorisations) and the real relations between the tertiary and other industrial sectors.

It is apparent upon closer examination that service occupations, defined as those the outputs of which are nonmaterial or ephemeral (Gershuny and Miles, 1983: 47) – are not limited to the service sector. An accountant working in a bank or in an electronics factory can be categorised as belonging to either the service or the manufacturing sector, though the work done may scarcely differ. Similarly, a carpenter working in a college of education or on a building site can be in either category. What this implies is that industrial classifications do not illuminate effectively the type of work performed, and that many producers of goods can be found in the service sector while many non-producers are in the primary and secondary sectors. In fact, Gershuny and Miles calculate that as much as half the growth in service occupations is a result of 'intra-sector tertiarization' rather than inter-sector shifts (ibid.: 125).

For example, when a manufacturer expands white-collar staff, perhaps in marketing, training or personnel, the firm is taking on service workers to help the company to stay in business more effectively, by for instance improving sales methods, teaching workers to be more efficient, or more carefully selecting employees. Each of these is an expression of an increased *division of labour* within a particular sector which boosts the number of service occupations. Most importantly, however, such examples must lead us to reject Bell's presentation of the service sector as some sort of parasite on the industrial base. If we can recognise similar occupations across the sectors (managers of all sorts, clerks, lawyers, etc.), then we surely cannot assert that in one sector some of these occupations are productive while in another all they do is consume the resources generated from the other. One has to cast doubt on the value of a sectoral division which suggests that one is wholly productive while the other is concerned only with consumption.

This does bring into question the use of regarding society in terms of separate sectoral levels, but the definitive rejection of such a way of seeing comes when one looks more closely at the service sector itself. What one sees there is that a good deal of service sector work is engaged, not in consuming the wealth created by industry, but in assisting its generation. Gershuny, in contending that 'the growth of the service sector of employment. . . . is largely a manifestation of the process of the division of labour' (1978: 92), leads one to realise the 'systematic link between the secondary and tertiary sectors' (Kumar, 1978: 204) and the consequent absurdity of sharply distinguishing realms in the manner of Bell. Browning and Singelmann, for instance, identify 'producer services' such as banking and insurance that are largely a 'reflection of the increasing division of labour' (Browning and Singelmann, 1978: 30). It is only by donning theoretical blinkers that one can perceive services as distinctly apart from production activities. The following observation from Gershuny is subversive of all theorisations which foresee services springing from the 'productivity' of the 'goods producing sector':

> the important thing to note about tertiary industry is that though it does not directly produce material goods, a large proportion of it is closely connected with the process of production in the slightly wider sense. The distribution

industry, for instance, does not itself make any material object, and yet is an integral part of the process of making things – if products cannot be sold they will not be produced. Similarly, the major part of finance and insurance is taken up with facilitating the production or purchase of goods. . . . though, in 1971, nearly half of the working population were employed in tertiary industry, less than a quarter of it – 23.1% – was involved in providing for the final consumption of services.

(Gershuny, 1977: 109–110)

The bald point is that the division of society into wealth-creating and wealth-consuming sectors or, more explicitly with Bell's theory of 'post-industrialism', into goods-'producing and service sectors, is an 'heroic oversimplification' (Perkin, 1990: 501). It feeds commonsense prejudices to think in these ways, but as historian Harold Perkin bitingly observes with reference to a closely cognate opposition:

The notion expressed by so many corporate executives, that the private sector produces the wealth which the public sector squanders, is manifestly false. It is just as valid to claim that the public sector produces and maintains, through the education and health services, most of the skills on which the private sector depends. In a complex interdependent society such claims and counter-claims are as naive and unhelpful as the pot calling the kettle black.

(ibid.: 502)

SERVICES AND MANUFACTURE

So the notion that services are readily separable from other work activities, let alone employment sectors, is false. It is possible to extend the critique by further drawing on the work of Gershuny, this time his collaborative research with Ian Miles. In a number of propositions developed in their book *The New Service Economy* (1983), Gershuny and Miles remind us of the *ex post facto* logic Daniel Bell draws upon to explain the growth of service sector employment.

Bell, starting from the indisputable fact that there is more service employment about nowadays, looks back from this conclusion to deduce its expansion from the rule that, as one gets wealthier, so one's additional income is spent on services. People must be spending more on services, argues Bell, since there are so many more service employees now. This does appear to be plausible. But it is mistaken, and it is a mistake which stems from Bell's failure to look at *what service workers actually do*. A great deal of service work can be accounted for by differentiation in the division of labour aimed at making more effective the production of goods.

Another major problem with Bell's account is his failure to consider that people might satisfy their service requirements by investing in goods rather than in employing service workers to do it for them. Gershuny and Miles come to this proposition by reversing Engels' theorem, wondering whether rather than increased riches leading to extra expenditure on personal services to satisfy needs, a relative increase in the cost of service workers, along with cheapened service products

44

becoming available, might have led to the satisfaction of service requirements by the purchase of goods rather than the employment of people. Put more directly: people want services as their standard of living increases, but they are not prepared to pay the price of people doing the services for them when there are service products available on the market that they can buy and use to do the service for themselves – for example, people want a convenient way of cleaning their homes, but because they are not prepared to pay wages to a cleaner, they get a Hoover to do it for themselves; or they would like their home decorated regularly, but because they will not pay for commercial painters, they invest in the DIY equipment and get on with it themselves.

Gershuny and Miles agree that people do want services, but the cost of having that service performed by another person becomes unattractive when set against the price of buying a machine to do it. In turn, this consumer demand for services in the form of goods 'can . . . produce pressure for innovation in service provision' (Gershuny and Miles, 1983: 42) which means that service requirements *impact on manufacture itself.* Instances such as the automobile industry and consumer electronics are pointers to the trend of fulfilment of service needs by goods rather than through employment of service workers. Gershuny himself claims, with impressive empirical documentation, that the spread of service products signifies the growth of a 'self-service economy' – almost the antithesis of Bell's 'post-industrial service society' (Gershuny, 1978: 81) – which is likely to continue to intrude into both service sector and service occupation employment. As he puts it:

> careful examination of changes in employment and consumption patterns . . . over the last 25 years reveals, not the gradual emergence of a 'service economy', but its precise opposite. Where we would expect, according to . . . [Bell's] dogma, to find a considerable rise in the consumption of services, we find instead a remarkable fall in service consumption as a proportion of the total. Instead of buying services, households seem increasingly to be buying – in effect investing in – durable goods which allow final consumers to produce services for themselves.
>
> (Gershuny, 1978: 8)

Furthermore, these service products 'form a fundamentally important source of change in the overall industrial structure' (Gershuny and Miles, 1983: 121). The 'industrialisation of service production' (p.84) is a pointer to what others whom we shall encounter in this book have called 'consumer capitalism', where the production and consumption of goods and services are to be regarded as intimately connected. And they underscore a recurrent criticism of Daniel Bell's theoretical and methodological presuppositions: that to conceive of society as divisible into distinctly separate realms is nonsense. The historical record shows that 'the economies of the Western world during the 1950s and the 1960s were dominated by the consequences of social and technological innovations in the nature of provision for a particular range of service functions, namely transport, domestic services, entertainment' (Gershuny and Miles, 1983: p.121). Far from the 'industrial' sector of the post-war societies determining the amount of wealth (or 'goods') available to pay for more service workers, *the major activity of industry was the manufacture of service products, in response to clear demand from consumers, that could substitute for service employees.* Bell's theorisation cannot begin to account for this, since an adequate explanation must jettison insistence on separate realms of society from the outset.

Gershuny's critique must mean that we reject Bell's notion of 'post-industrial' society. And this rejection must be quite sweeping, dismissing everything from Bell's anti-holistic mantra (societies are *not* radically disjuncted, but intricately connected) to his general account of social change as an evolution through stages towards a 'service economy'. His explanation for the emergence of PIS is misconceived, his description of an emergent 'caring' society unconvincing, and his insistence that it is possible to identify separate employment sectors (which are yet causally connected, with services being dependent on the goods-producing level) is incorrect.

One is forced to take the view that more service sector employment, more white-collar work, and even more professional occupations – all of which Bell correctly highlights – do not announce a 'post-industrial' epoch. On the contrary, each of these trends is explicable as an aspect of the continuity of an established, and interdependent, socio-economic system. Furthermore, while these shifts and changes do lead to increases in information and information activities, it is an error to move from this to assert that a 'post-industrial information society' has emerged.

THEORETICAL KNOWLEDGE

I have shown that the foundations of Bell's 'post-industrial' model are insecure. In so doing it has become apparent that his equation of 'post-industrial' and 'information' societies is untenable: since his argument that professional, white-collar and service work represents PIS is misconstrued so his assertion that 'post-industrialism' is an adequate account of the 'information age' must collapse. Above all perhaps, there are no signs that a *break* with former societies is appearing – indeed quite the reverse. As the excellent Krishan Kumar observes, 'the trends singled out by the post-industrial theorists are extrapolations, intensifications, and clarifications of tendencies which were apparent from the very birth of industrialism' (Kumar, 1978: 232). So we must refuse the idea of 'post-industrialism' as a way of understanding present concerns with information. This does leave us with the undeniable fact that a good deal more information work is taking place in advanced societies, though it is insufficient to assert that this in and of itself engenders a new sort of society. Just as one cannot assert that more service occupations prove there is emerging a new sort of society, so it is not enough to contend that *more* information of itself represents a new society.

However, if we cannot accept that more information can of itself create a new sort of society in the way Bell envisages, there are other elements of his views on information which deserve attention.

Describing 'post-industrial' society, Bell sees not only an expansion in information as a result of more service sector employees. There is another, more qualitatively distinct, feature of information in PIS. This is Bell's identification, as an 'axial principle' of the society, of what he calls 'theoretical knowledge'. Now, while an expansion of professionals will certainly increase the number of people using and contributing to 'theoretical knowledge', we are not considering here a merely quantitative – and hence relatively easily measured (numbers of lawyers, scientists and so forth) – phenomenon. It is, rather, a feature of PIS which distinctively marks it off from all other regimes and which has profound consequences. It is not even altogether clear how it fits with much of Bell's other descriptions of PIS (occupational changes, sectoral shifts and the like), since the centrality of 'theoretical knowledge' to PIS does not, in principle at least, require major changes in jobs or, indeed, in the nature of work.

It does, however, have enormously significant effects on all aspects of life. Bell's argument is that 'what is radically new today is the codification of theoretical knowledge and its centrality for innovation, both of new knowledge and for economic goods and services' (Bell, 1989: 189). This feature allows Bell to depict 'the post-industrial society . . . [as] a knowledge society . . . [because] the sources of innovation are increasingly derivative from research and development (and more directly, there is a new relation between science and technology because of the centrality of *theoretical* [*sic*] knowledge') (Bell, 1976a: 212).

The constituents of 'theoretical knowledge' can be better understood by contrasting PIS with 'industrial' society. In the past, innovations were made, on the whole, by 'talented amateurs' who, encountering a practical problem, worked in an empirical and trial-and-error way towards a solution. One thinks, for example, of George Stephenson developing the railway engine: he was faced by the practical difficulty of transporting coal from easily accessible collieries situated a distance from rivers – and in response he invented the train which ran on tracks and was powered by steam.

In contrast, PIS is characterised by 'the primacy of theory over empiricism and the codification of knowledge into abstract systems of symbols that . . . can be used to illuminate many different and varied areas of experience' (Bell, 1976a: 20). This means that innovation nowadays is premissed on known theoretical principles – for example computer science takes off from Alan Turing's seminal paper (On Computable Numbers) (1937) which set out principles of binary mathematics, and the extraordinary miniaturisation of integrated circuits that has allowed the 'microelectronics revolution', was founded on known principles of physics.

The proposal is that nowadays theory is pre-eminent not just in the area of technological innovation, but even in social and economic affairs. For example, governments introduce policies that are premissed on theoretical models of the economy. These may be variable – Keynesian, monetarist, supply side and so forth – but they are, nonetheless, theoretical frameworks which underpin any day-to-day decisions ministers may make in response to exigencies. Elsewhere, one may instance the primacy of theory in social affairs, for instance in the creation of educational and medical provision, where experts make their decisions on the basis of theoretical models of the operation of family structures, lifestyle variations and demographic trends. Undeniably the theoretical knowledge used here is often imprecise, but this in no way undermines the point that it is a prerequisite of action. The truth is that, where once actions were responsive to practical issues (a technical problem, a social obstacle), nowadays much of life is organised on the basis of theories – abstract, generalisable principles – of behaviour.

Bell thinks this change has important consequences. Perhaps most important, the primacy of theory in all spheres gives PIS a capacity to plan and hence to control futures to a much greater degree than previous societies. This capability of course accords with the professionals' predisposition to organise and arrange life. In addition, theories are made more versatile by to the advent of information technologies. Computerisation allows not just the management of 'organised complexity', but also, through programming, the creation of 'intellectual technology' (Bell, 1976a: 29) which incorporates knowledge (rules, procedures and the like) and in turn facilitates innovations based on theoretical knowledge.

This 'primacy of theoretical knowledge' is an arresting idea, one which, in reversing the very principles of organisation and change prevalent in 'industrial' society, establishes *prima facie* a definition of a new type of society hinging on information/knowledge. We seem to be talking here not about more white-collar work or more bits of information being produced, but of an entirely new foundational principle of social life.

There are, however, several objections to Bell's contention. One queries the extent to which 'theoretical knowledge' really is novel. Anthony Giddens, for example, argues that 'there is nothing which is specifically new in the application of 'theoretical knowledge' to productive technique. Indeed, as Weber stressed above all, rationality of technique. . . . is the primary factor which from the beginning has distinguished industrialism from all preceding forms of social order' (Giddens, 1981: 262). This being so, PIS's emphasis on knowledge is essentially an extension and acceleration of industrialism's priorities and we are back to rehearsing doubts about the novelty of PIS.

But Giddens' objection does beg a further question: just what, precisely, does Bell mean by 'theoretical knowledge'? Weber's conception of formal rationality underpinning purposive action, referred to by Giddens, could apply on one definition – after all it involves abstract and codifiable principles, rules and regulations (the entire bureaucratic machine), as well as requiring from participants command of abstract knowledge (how the system works). As such, this could satisfy as a description of the work of a great number of professional occupations such as law, accountancy and even medicine. Bell scarcely makes clear whether this definition would be included in his notion of 'theoretical knowledge'. Indeed, he is noticeably vague when it comes to the question of stipulating what it constitutes.

Yet it is essential, if 'theoretical knowledge' is to be taken as a defining criterion of PIS, that we are told what it is. Unavoidably, this lack of clarity over its meaning leads to major criticisms being levelled at Bell. For instance, if the 'primacy of theoretical knowledge' is taken to refer to known scientific principles (the boiling point of water, the conductivity of elements, chemical reactions, etc.) which are codified in texts of one sort or another, then this is one matter. However, if 'theory' is interpreted to mean theoretical models or even hypotheses such as the relation between inflation and employment, then this surely is quite another. Again, if the 'primacy of theoretical knowledge' is taken to mean the privileged position in innovation of research and development investments and teams, then this is another still. Finally, if 'theory' is perceived to be the prominence of expert systems in the organisation of modern life (the electricity grid, waste disposal, air traffic control or television and telephone networks), then again quite a different definition is being introduced. And the problem with Professor Bell is that he uses each and all of these definitions in his musings on the 'primacy of theoretical knowledge', making it exceedingly difficult to pin him down to anything so mundane as an empirical test.

These ambiguities are not trivial matters either, since, when it comes to attempts at gauging the rise of 'theoretical knowledge' and interpreting the significance of what has been measured, Bell lands in serious difficulties. For instance, counting the growth of educational certificates as an index of the rising curve of theoretical knowledge (which Bell does) is extremely dubious. Accumulation of school and college diplomas can scarcely be said to indicate the emergence of 'theoretical knowledge' when one reflects on criticisms such as the inflation of qualifications, their connectedness to the substantive realm of employment, and widespread scepticism about the standards achieved in much education.

If we examine research and development (R&D) budgets and organisation as signs of the growth of theory's primacy (which Bell does: 1976a: 250–262), we realise quickly that there are big problems: (a) with the presumption that R&D projects serve to heighten the import of theory – for the most part R&D is applied rather than basic in orientation, aimed at making practical improvements in existing products rather than developing new knowledge (Freeman, 1974, 1987); and (b) with any suggestion that the 'theoretical knowledge' of the R&D team is decisive – quite the reverse seems to be the case, with corporate leaders establishing R&D budgets and directions with a firm eye to commercial markets or state contracts. It cannot be denied, of course, that R&D budgets have risen, but there is convincing evidence that researchers continue to act as 'servants of capital' or of state paymasters which supply investment in R&D with clear priorities in mind (Robins and Webster, 1989: 237–243). In short, the subordination of R&D programmes to economic and political powerholders implies the continuation of existing social relations rather than the emergence of a new 'post-industrial' society in which 'theory' is decisive.

Finally, were one to consider the 'primacy of theoretical knowledge' to be evidenced in the prominence in contemporary affairs of complex organisations which require scientific expertise to run them (there are many examples, from large organisations such as hospitals and universities to institutions like the rail and road networks), then one really must encounter the charge – recurrently levelled at Bell – of technocracy. That is, in presuming that technological knowledge empowers so much as to mean it is 'axial' in PIS, one confuses indispensability with power. It is all very well to argue that experts are essential to the world in which we live, but it is a fatal flaw to believe that this vaults them to a primary role. Just because we rely on experts, even those with high levels of esoteric knowledge, to supply us with things like potable water and continuous electricity, we must not presuppose that this endows them – or the 'theoretical knowledge' they possess – with power over the rest of society.

CONCLUSION

Daniel Bell began some years ago to substitute the concept 'information society' for 'post-industrialism'. But in doing so he did not significantly change his terms of analysis: to all intents and purposes, his 'information society' is the same as his 'post-industrialism'. However, we have seen in this chapter that his analysis just cannot be sustained.

Undeniably, information and knowledge – and all the technological systems that accompany the 'information explosion' – have quantitatively expanded. It can also be readily admitted that these have become central to the day-to-day conduct of life in contemporary societies. Nonetheless, what cannot be seen is any convincing evidence or argument for the view that all this signals a new type of society, 'post-industrialism', which distinguishes the present sharply from the past. To the extent that this criticism is valid, then all talk of developments in the informational domain representing the coming of 'post-industrial society' must be refused.

It has been demonstrated that Daniel Bell's division of society into separate realms, and his further division of the economy into distinct employment sectors, collapses on closer examination. Services, white-collar work, even professional occupations have grown, and they have all manifested greater concern with handling, storing and processing information, but there is no reason here for interpreting their expansion as consequent upon more wealth flowing from a 'goods producing' sector to a separate realm of consumption. On the contrary, services have expanded to perpetuate and secure an established, interconnected, economy (and, indeed, wider political and cultural relations). *There is no novel, 'post-industrial' society: the growth of service occupations and associated developments highlight the continuities of the present with the past.*

Further, Daniel Bell's emphasis on 'theoretical knowledge', analytically if not substantively separable from the more quantitative changes referred to above, has an initial appeal. Being a qualitative change, with profound consequences for planning and control of social affairs, it is an arresting thought for anyone interested in social change and the possible significance of information/knowledge in the contemporary world. However, it too cannot be sustained, since in the writing of

Bell it is either too vague to be applicable or, where made more precise, serious doubts may be cast on its novelty and weight.

We remain with the fact of living in a world in which information and informational activity forms an essential part in daily organisation and much labour. On any measure the scale and scope of information has accelerated dramatically. Understandably, social scientists yearn to explain and account for this development. Our conclusion here is that it cannot be interpreted in Professor Bell's 'post-industrial' terminology. Bell's ambition to impose the title 'post-industrialism' on to the 'information society' simply will not do. If we want to understand the spread and significance of information in the present age we must look elsewhere.

4

INFORMATION, THE NATION STATE AND SURVEILLANCE: ANTHONY GIDDENS

Anthony Giddens (born 1938) is unarguably the most significant social theorist in post-war Britain, at the very forefront of producing an 'outstanding indigenous sociology' (Anderson, 1990: 52). Giddens' ambition is both to recast social theory and to re-examine our understanding of the development and trajectory of 'modernity'. It is not the aim of this chapter to assess Giddens' *oeuvre*. Instead, what I should like to do is take from his book *The Nation State and Violence* (1985) several ideas which I think help us to see the development and significance of information in an illuminating way. I had better say also that I intend to take off from some of Giddens' work since, from a review of several of his ideas, I shall extend freely into other authors' views and empirical work which he himself leaves alone.

Anthony Giddens does not write much about the 'information society'. It is not a concern of his to discuss the status of this particular concept, not least because he would surely be sceptical of the proposition that we have recently seen the emergence of this new type of society. Indeed, he has quite directly asserted that 'although it is commonly supposed that we are only now in the late twentieth century entering the era of information, modern societies have been "information societies" since their beginnings' (Giddens, 1987: 27). Consonant with this statement, Giddens' theorisation leads one to argue that the heightened importance of information has deep historical roots, so deep that, while one may concede that information today – in an era of what he calls 'high modernity' – has a special significance, it is not sufficient to mark a system break of the kind Daniel Bell conceives as 'post-industrialism'.

Central to Giddens' concerns is critical engagement with classical social theorists, most notably Karl Marx, Emile Durkheim and Max Weber. And his aim, like theirs, is to understand the cluster of changes termed the emergence of 'modernity'. However, Giddens finds Marx's explanation – the dynamics of 'capitalism' – and the Durkheimian and Weberian master keys – the notions of 'industrialism' and 'rationalisation' – inadequate. There are other contributory factors in the making of the modern world which the great tradition either understated or overlooked. More particularly, in *The Nation State and Violence* Giddens emphasises two associated features of modernity underplayed by the classical thinkers. These are the significance of *heightened surveillance* in the genesis of modernity, and the import of *violence, war and the nation state* in the development of the contemporary world. Giddens' endeavour to illuminate these offers us a means of presenting an interesting perspective on the origins, significance and development of information.

ORGANISATION AND OBSERVATION

Before we get on to direct discussion of Giddens' ideas it is worth establishing a point which is preliminary to all that follows. This is that the world in which we live is much more *organised* than ever before: our everyday lives are planned and arranged by institutions in unprecedented ways.

This is not to imply some inexorable decline in personal freedom. There can be no doubt that in the past circumstances massively restricted humankind: hunger, the uncertainties of nature, the impositions on women of multiple pregnancies. Above all, the dull compulsion of everyday existence must have imposed enormous limitations on people besides which most modern constraints can seem scarcely significant. The premiss that life today is more routinely and systematically managed does not mean that nowadays we inhabit some sort of prison. Indeed, as will become clear, our increased liberties are very often closely correlated with greater organisation – though, of course, this does not have to be the case. An obvious counter-example would be the Holocaust, an event inconceivable 'without modern civilization and its most central essential achievements' of 'rational, carefully calculated design' (Bauman, 1989: 87, 90), devoted in that case to systematic genocide.

Life nowadays is much more methodically arranged than ever before. This has become about not least because of modern capacities to limit the traditional constrictions of nature. As we have become able, for example, to dispose hygieni-cally of human waste and to create plentiful supplies of food, so life has moved from governance by nature to organisation by elaborate social institutions. And here are instances whereby increased options for people (reliable sanitation and sufficient food) have accompanied, and indeed been premissed upon, the develop-ment of organisational structures.

A moment's thought will bring home the extent of modern-day organisation. Consider the school system in Britain, an astonishing organisational accomplish-ment which brings together thousands upon thousands of teachers and pupils at pre-ordained times, to undertake pre-established activities which, if locally vari-able, have a great deal in common across the nation, and all of which is arranged to ensure continuity over the years. Or consider the astonishing organisational arrangements that lie behind an activity essential to all of us – shopping for food. The daily routine of co-ordinating suppliers, producers, manufacturers, transport and customers that is required of today's supermarkets (typically stocking some-thing like 16,000 different items, many of which are perishable, thereby compounding problems for the retailer) is a spectacular organisational achievement compared to previous ages.

We ought to emphasise that this organisation is often extremely complex.

Consider, for instance, the planning that is a prerequisite of train and bus schedules, of the electricity supply industry, of television programming, of credit card systems, or of the production of clothing for large retail outlets – or even something as mundane as the cereals most of us eat at the breakfast table. It matters neither that we reflect little on the 'abstract' and 'expert' systems (Giddens, 1991) which handle these arrangements nor that, for the most part, we have 'trust' (ibid.) in their reliability. The fact remains that modern life is unprecedentedly socially organised.

All of this is to spout sociological truisms. However, one consequence is easily overlooked, though it will be a major theme of this chapter: that to organise life information must be systematically gathered on people and their activities. We must *know about people* if we are to arrange social life accordingly: what they buy, when and where; how much energy they require, where and at what times; how many people there are in a given area, of what gender, age and state of health; what tastes, lifestyles and spending capacities given sectors of the populations enjoy. . . . Bluntly, *routine surveillance* is a prerequisite of effective social organisation. Not surprisingly, it is not difficult to trace the expansion of ways of observing people (from the census to check-out tills, from medical records to telephone accounts, from bank statements to school records), moving in tandem with the increased organisation which is so much a feature of life today. *Organisation and observation are Siamese twins, which have grown together with the development of the modern world.*

Arguments that we live a much more organised existence are by no means new in social science. They are evident in Theodor Adorno's ruminations on the 'administered society', in Michel Foucault's documentation of the 'carceral networks' characteristic of contemporary existence, and they are present too in Max Weber's depiction of the 'bureaucratisation' of the modern world. But it is striking that the informational dimensions of this development have been understated. Not in Foucault certainly, but in Weber's writing on the relentless spread of calculation which finds expression in bureaucratic apparatuses it is noticeable how the pivotal role of information is neglected. When one considers that, to Weber, bureaucracies develop to manage affairs in more technically efficient ways, it is surprising that neither he nor his followers expanded much on the informational consequences of systematic book-keeping, the building up of records, the documentation of cases, the codification of rules and procedures, all aimed at facilitating a maximally organised existence. As Chris Dandeker cogently observes, 'the age of bureaucracy is also the era of the information society' (Dandeker, 1990: 2).

PARADOXES OF MODERNITY

It is well known that most commentators have had decidedly glum opinions about this. Adorno's pessimism is famous. As for Foucault, his own disclaimers to the contrary, it is hard to avoid the conclusion that 'disciplines and surveillance become increasingly . . . more detailed, more complex, more efficacious and more determinant' (Turner, 1992: 131). And Weber's resignation to the inevitability of bureaucratisation did not lessen his gloom about the loss of 'heroes', his mourning for the 'departure from an age of full and beautiful humanity', his Nietzschean horror at the prospect of a world filled with 'specialists without spirit, sensualists without heart', and his profound distaste for the 'mechanized petrification' which accompanies the 'iron cage' of rational-legal organisation (Weber, 1976: 181–182).

Given the wide currency of such views – amounting in all essentials to what we nowadays call Orwellian images of society – one may comment, with the informational elements of organisations to the forefront of attention, on what may be referred to as paradoxes of modernity.

It is useful to distinguish *individuation* from *individuality*. The former refers to the situation when each and every person is known about, hence identified by a singular record, say of name, date of birth, residency, employment history, educational achievements and lifestyle preferences. The latter, which so many commentators believe to be threatened by increased social organisation and the observation which is its accompaniment, is about being in charge of one's own destiny, having genuine choices in and control over one's life – things inimicable, it would appear, to intrusive institutions and their information-gathering impulses.

Frequently individuation and individuality are conjoined, with the undoubted increases in individuation being taken to mean there has been a decline in individuality. Now it is undeniable that individuation requires that people be monitored and observed, but the development of files on individuals documenting their earnings, housing circumstances and the like may in fact be requisites of enhancing their individuality in so far as this relies on their being treated as particular beings and, let us say, being sure of receiving entitlements without which they may be limited in their capacity to be true to themselves, to have the possibility of exercising genuine choices and so on. Put in less abstract terms, if we as a society are going to respect and support the individuality of members, then a requisite may be that we know a great deal about them. For instance, if each of us, as an individual, is to have a vote, then we must be individuated at least by name, age and address. Again, if as a society we consider that members must reach a certain level of housing provision and material sufficiency in order to fulfil their individuality (if people are cold, alone and living in abject poverty there is widespread agreement that their individuality is thwarted), then it is an essential requisite of meeting those needs that we individuate people and detail their precise circumstances. Nicholas Abercrombie and colleagues (1986) acknowledge this paradoxical situation when they write that 'individuation, by enhancing the rights of the individual, leads to greater surveillance and control of large populations' (p.155), but they underline the positive dimensions of this process:

> Rather than seeing passports, insurance numbers, consumer credit-cards and other features of regimentation as inevitable forms of surveillance and domination, we can see such standardization of persons in a more positive

light: namely, as unavoidable consequences of the requirement to treat people equally.

(ibid.: 154)

This point may be taken further, beyond the idea that information needs to be gathered in order that people may gain entitlements. It is clear that in many spheres monitoring of individuals is a foundation for the operation of complex organisations which, through the services they supply, can enhance the individuality of customers. For instance, the telephone network individuates every user with great precision and accumulates a massive amount of detail about each subscriber (all users have a unique number and every call is automatically logged for destination and duration). Upon the basis of this information are established telecommunications networks which extend into most homes in advanced societies and reach out across the globe. For those people with appropriate connections these organisations offer enormous enhancements to their lives. At the touch of a button people may keep up friendships, family and professional relationships, links which clearly enhance the sense of self and individuality. Much the same point can be made about the construction of banking networks. Most people nowadays have credit cards of one sort or another through which each and every transaction made may be recorded and an individuated profile of spending patterns constructed (the bill). But if it is on the routine monitoring of an individual's purchases and payments that complex banking networks operate, then these very processes can increase the individuality of actors by making credit and the transactions of everyday life considerably easier. Anyone who has tried to book a hotel or hire a car or even travel without excessive fear of theft and anxiety about handling foreign currencies will appreciate this point.

If we cannot therefore straightforwardly equate greater information about people with a diminishment of individuality, there is yet another paradox that requires comment. This stems from recognition that we have emerged from a world of neighbours and entered what has increasingly become one of strangers. Here we have the old theme in social science of a shift from *community* (crudely, the familiar, interpersonal and village-centred life of pre-industrialism) to *associations* which involve the mixing of people unknown to one another save in specific ways such as bus conductor, shop assistant, and newsvendor (crudely, the urban-oriented way of life of the modern). Ever since at least Simmel we have appreciated how disorienting and also often liberating the transfer from closed community to a world of strangers can be. The city may fragment and depersonalise, but in doing so it can also release one from the strictures of village life. With the shift towards town life comes about a decline in personal observation by neighbours and, accompanying this, a weakening of the power of community controls that are exercised on an interpersonal basis. Entering urban-industrial life from a country existence one is freed from the intrusions of local gossip, of face-to-face interactions, from close scrutiny of one's everyday behaviour by neighbours. . . . By the same token, in the urban realm one can readily choose freedom, to be as private as one likes, to mix with others on one's own terms, to indulge in the exotic without fear of reprimand, to be anonymous. . . .

The paradox here is that urban societies – and advanced societies are overwhelmingly urban – in being much more socially organised than communal-based modes of life, must gather extremely detailed knowledge about their publics in order to function. And in key respects the information gathered by these institutions is more detailed, more insinuating, more precise and more individuated than anything garnered in a pre-industrial community. There talk and memory would be a major means of gathering and storing information; today the information is put together and stored by a variety of means (computerised records, merged data bases, written records, the routine 'metering' of actions such as use of electricity or banking services) and accumulated through time. Anyone doubtful of the precision or weight of such information might reflect on the tales a few months' supply of bank or credit card statements could tell about them. Or one might consider the intimate profiles that one's medical, educational or employment histories reveal.

The modern, impersonal life of associations between strangers nevertheless can entail the collection of even greater information/knowledge about individuals than the closed, personal world of neighbours. It may be that we can readily shed the cloying grip of family and friends in the city, but we can scarcely avoid the surveillance of the Inland Revenue, medical services or local authority. Of course, we need to add the important qualification that the organisations which observe us today are often impersonal (the tax office, the housing department, the hospital), and the information they harvest is as a rule discrete and kept that way, factors of considerable importance to those worried by the threats to civil liberties posed by increased surveillance.

In addition, much of the observation undertaken today is of an anonymous kind, by which I mean that a good deal is known about people's lives – their shopping preferences, their sexual proclivities, their lifestyles, their political allegiances – but, intimate though it often is, it does not name, still less individuate, the subjects who supply the information. An upshot of this is that people are most closely observed nowadays, so much so that, living amidst strangers, they remain much more intimately known than any previous generation, even those living in a cloistered community. Today we know a great deal about people's sexualities, about their aspirations and secret desires, and also about political preferences at a given time. All such information sets the contemporary society well apart from pre-industrial England when mechanisms for gathering such information were not in place. However, the information gathered about ourselves and others, which feeds into each of our own perceptions and even behaviour, does not usually reach to the level of identifying the individuals from whom the original data was gleaned.

In spite of this, the information so gathered is frequently essential for the functioning of modern organisations (political parties, retail companies, family planners etc.) and, moreover, it very often feeds back to other individuals (through media and educational institutions especially) who, having learned more about people and expectations, are themselves better equipped to make choices about the conduct of their own lives (e.g. about the range of lifestyles available in society at any given time, about different sexual preferences, about the variety of child-rearing practices). Again we encounter the paradox: as more is known about people, so individuals may get opportunities to enhance their own individuality by making 'choices' of their own.

In what follows it is as well to bear in mind these observations because, when it comes to examining the growth of surveillance, it is easy to believe that more observation intrudes upon the liberties of individuals, just as greater organisation appears, necessarily, to diminish the individual's autonomy. In such circumstances the readily available judgement – how awful! – may well be an oversimplification. When it comes to analysis of the state's role in organisation and observation, something with which this chapter is centrally concerned, such a judgement is especially appealing, which is yet further reason to beware impulsive judgements.

THE NATION STATE, VIOLENCE AND SURVEILLANCE

We may now turn more directly to the work of Anthony Giddens to understand better the particular forms of the expansion of routine surveillance and organisation in modern times. Giddens identifies a number of features germane to my concern with information. Most importantly, he pays special attention to the role of the state in the making of the modern world and in the expansion of surveillance.

It is important to preface my remarks with a point that Giddens has made many times. This is that, in most parlance, when we talk of 'society' we are actually referring to *nation states*. For example, when we study 'modern society' as a rule we study 'modern Britain' (if we are British) and, when we compare different 'societies', we generally contrast nation states (for instance Britain and America).

Now, while an equation of 'society' and 'nation states' may be satisfactory for everyday conversations and even for a good deal of sociology, it has to be recognised that the two terms are not synonymous. We need to acknowledge that the nation state is a particular kind of society, one created very recently in world history. If one spans little more than the past three centuries, one realises that most peoples have not lived in nation states, but in agrarian collectivities with loose and fluid frontiers. Where there were links between peoples these were in the form of religious commmunities such as Islam, Buddhism and Confucianism or in dynastic regimes like the House of Habsburg and that of the Bourbons (Anderson, 1991: 12–22).

The concept of a nation state came into being during the late seventeenth and eighteenth centuries and, while it has been at the centre of the construction of the world as we know it, and thereby has exercised a profound influence on our ways of seeing, it must be examined as an artifice. The nation state is *not* 'society', but a particular type of society that has distinctive characteristics.

Here I may telegraph a central theme of Giddens' argument. He contends that from the outset in the nation state, conceived as a bounded area over which is exercised political power, information has an especial significance. Indeed, from their establishment, nation states are 'information societies' in that they must, minimally, know their own members (and, necessarily, those who do not belong). Giddens believes that nation states must maintain hold of both 'allocative resources' (planning, administration) and 'authoritative resources' (power and control) and that, while these tend to converge in the modern state, a prerequisite of both is effective surveillance. It follows that:

> modern societies have been. . . . 'information societies' since their inception. There is a fundamental sense . . . in which all states have been 'information societies', since the generation of state power presumes reflexively gathering, storage, and control of information, applied to administrative ends. But in the nation state, with its peculiarly high degree of administrative unity, this is brought to a much higher pitch than ever before.
>
> (Giddens, 1985: 178)

But this is too abstract. What we need to do is to elaborate more of the detail of the argument that the nation state has a particular interest in and reliance upon information gathering and storage, and that this impulse has profound consequences. Essential to this task is to describe further some of the major features of the nation state.

First, *the modern world is constituted by nation states.* This is in no way to underestimate the process of what is now known as globalisation whereby national frontiers appear increasingly subordinate to financial and economic trends. I deal with some of these issues in a later chapter, but here the emphasis on the division of the world into nation states gives us a necessary sensitivity to vitally important features of contemporary life.

Amongst these is the fact that nation states are essential to many, perhaps most, people's identities. Charlie Chaplin's famous declaration that he was a 'citizen of the world' is arresting precisely because, to most people, it is so idiosyncratic. To the majority national allegiance ('I am British, French, German, American. . . .') is a central element of their being. The issue of national identity is extraordinarily complex and fraught – and at the core of a great deal of modern political movements. At one pole, whether one watches one's national football team on television and wills them to win or roots for one's country's representatives at the Olympics, there is evidence of national consciousness of some sort. At another, we have expressions of nationalism which are autocratic, racist and belligerent – the 'ethnic cleansing' pursued in former Yugoslavia is a reminder of just how virulent this can be. But everywhere, to a greater or lesser degree, nation states influence identities by constructing mythic pasts made up of legends and literature, traditions and celebrations, customs and caricatures. Study of these 'collective identities' (Schlesinger, 1991) has produced a voluminous literature, all of which agrees that they are an inescapable feature of modernity (Smith, 1986).

The 'imagined communities' (Anderson, 1991) which make up national identities, it needs to be stressed, have no natural foundation, however much myth-makers proclaim, in Britain, our 'Anglo-Saxon heritage', the 'timelessness of the monarchy', or 'our yeoman traditions'. Indeed, the more one examines national identity, the more one uncovers ill-thought-out, irrational and contradictory beliefs which switch between vague national symbols (Hyde Park Corner, the Palace of Westminster), an imaginary past (Robin Hood, King Arthur, the Norman yoke), and evocative regional scenes ('true England' can be the Sussex Downs, the Dorset hills or Durham Cathedral).

But however much academic analysis may cast doubt on the veracity of 'national identity' in its cultural forms, the fact remains that it has had an extraordinary potency in modern history. As many a Marxist has reluctantly had to concede, the masses have wrapped themselves in their national flags with much more alacrity than they have followed the call of 'workers of the world unite'. Moreover, in defining who belongs to a particular nation, there is necessarily a definition of who does *not* belong. While this may not be too much of a problem when it comes to legal conceptions of nationality (i.e. who is to carry a passport and have access to other citizenship rights), in the realm of the cultural – that area of feelings, meanings, identities – it is paramount. As a result, it is with good reason that Stuart Hall fears a 'closed and exclusive definition of 'Englishness' . . . [which] functions as the coded language for colour' (Hall, 1992: 6). When Englishness is imagined as a cultural identity which, if variegated (cricket on the green, real ale, regional dialects, football), constantly holds to the notion that 'whiteness' is a requisite of inclusion, then for ethnic minorities of colour there are most serious problems and threats.

It is not surprising that the nation state remains quite central to people's identities when one notes that the emergence of modernity has been experienced within a context of developing and consolidating nation states, most obviously in Europe and of course the United States. The orthodoxy amongst social theorists was that the nation state and associated nationalisms would irrevocably decline when faced with the logics of 'industrial' or 'capitalist' expansion, but this simply has not been so. In fact, as we shall see, much of the dynamism of industrial capitalism has come from the imperatives of the nation state itself, something that in turn stimulates feelings of national consciousness.

Furthermore, the nation state remains absolutely crucial to a very great deal of economic, social, cultural and legal life. One has but to reflect on the significance of fiscal policies and educational strategies, or the complex issues surrounding law and order to appreciate this and to better understand the continued salience of the nation state in people's lives.

At the same time, it is sobering to be reminded of the novelty of the nation state. So many of us have become so accustomed to the state's presence, institutional and ideological, that it can appear to have an extraordinary permanence. However, even 'traditional' nation states are little more than a couple of centuries old and, it should be stressed, none is fixed. Even the United Kingdom has a history of less than three hundred years, and still today there are recurrent challenges from Scottish, Welsh, and especially Northern Irish separatist constituencies. One has only to consider the recent events in Eastern Europe to understand the mutability of nation states:

the break-up of the USSR, the reunification of Germany, the destruction of Yugoslavia. Little more than a casual glance across Europe forcibly reminds us that there is scarcely a nation state which is not challenged by internal nationalisms; and a closer look at the Middle East reveals nation states (Yemen, Kuwait, Iraq, Jordan, Oman, Saudi Arabia) established in recent decades on societies which hitherto were tribal (cf. Friedman, 1990).

I lay emphasis on the primary importance of nation states to socio-economic organisation and identities alongside their novelty and tendency to recompose because this allows us to pay due attention to a second key feature of the nation state. This is that the overwhelming majority of nation states have been *created in conditions of war and all are sustained by possession of credible defence*. War and preparedness for war have been fundamental contributors to the nation state. Any analysis of British history makes the point forcefully enough: the Act of Union emerged from military defeat of the Celtic fringe, and important preconditions were that strong monarchs were able to defeat and place under their control once autonomous barons while offering some security from outside invasion. The more recent history of Britain, notably that of the days of Empire, illustrates dramatically the readiness of nation states to fight over territories and, by no means least, the contribution this made to national consciousness (one might recall here those maps of the world covered in 'British red' studied by all schoolchildren well into the 1960s, the resonance of Kipling's poetry, or the 'Argie bashing' sentiment unleashed in 1982 'when the drums began to roll').

This point may be put in a less dramatic way. From the definition of the nation state as having sovereignty over a given territory, it follows that a minimal responsibility of national governments is upholding the integrity of their borders. Most regimes invest heavily in defence to this end, but even the very few which do not aim to secure their frontiers by strategic alliances with states which can offer a credible deterrent to potential aggressors. *Preparedness for war* (i.e. a credible defence capability) *is a requisite of all nation states*, a principle that has repeatedly been put to the test throughout modern history.

A third key feature of the nation state is closely connected to the second. This is that *modern warfare/defence has become much more decisively implicated with the wider society*. On one level this simply means that greater proportions of the population are engulfed by modern warfare. Thus, through time, one can trace an increase in the number of casualties of war both amongst combatants and, still more strikingly, amongst the civilian population. Crudely, war kills and maims more people than ever before.

It is usual to see the 'Great War' (1914–1918) as marking a decisive turning point in the consequences of war, and undeniably the number of military casualties was unprecedented. However, though Britain lost almost ¾ million men then, 'the civilian population suffered little physically. Less than 1,500 civilians were killed by enemy action from sea or air' (Taylor, 1965: 121). As this century has unfolded it is amongst the civilian populations that war has wreaked its most severe damage.

61

Modern warfare leaves no hiding place from aerial and other forms of attack. The Second World War, though actual combatant losses were much lower for Britain than in the First World War, ended with over 45 million dead, the vast majority non-uniformed, losses amounting to around 10 per cent of the populations of Russia, Poland, Yugoslavia and Germany.

If modern wars between states have increased in ferocity in this sense there remains another, related, way in which warfare has extended deeper into the social fabric. In contemporary wars, the entire society is mobilised in support of the war effort. On the one hand, it involves 'digging for victory' and the substitution of female and young labour for men conscripted to fight. On the other hand, it emphasises the growing intimacy of industry and the war effort in the late twentieth century.

It has been persuasively argued that, over the long term, there is a positive correlation between a nation's industrial strength and its military effectiveness: 'the countries with the deepest purse ha[ve] prevailed in the end' (Kennedy, 1988: 458). An associated feature of this truism (accelerating since the opening decades of the century) has been a close connection between industrial activity and preparedness for war. As Giddens puts it, in observing the developing links between the state's war activities and industries such as chemicals, energy and engineering, it was during and after the First World War that commentators began to recognise 'the integration of large-scale science and technology as the principal medium of industrial (and military) advancement' (Giddens, 1985: 237).

It follows that, with war/defence being profoundly influenced by industry's capacity to produce equipment essential for its conduct, then what has been called the 'industrialisation of war' is a central feature of the modern world. This, in turn, means that the state's defence policies and spending habits have important consequences for the industrialisation process, though, as I have suggested, classical social theory tended to understate its significance, preferring the view that industrialisation would make the nation state – and the cognate requirement of credible defence – outdated.

This is not the occasion to elaborate on the detail of the connections between the state, military needs and industrialism (see Robins and Webster, 1989: 233–255). What I do want to do is focus on the linkages between information, information technologies and the state/military/industry nexus. Having observed that, in an epoch of total war, industry and defence become closely connected, there is no difficulty in establishing that information and information technologies have an especially intimate relation with the waging of and preparation for war. The reasons are obvious: modern warfare/defence must be maximally credible, to which end it must be based on the leading technologies. Not surprisingly, a great deal of research and development expenditure comes from the state (in Britain approximately half of total funds), and much of this is dedicated to supplying state-of-the-art military technologies (in Britain 50 per cent of the state's R&D spending is to this end). Furthermore, there cannot be a more leading edge of military technology than that of electronics, which 'today forms the underpinning of national security systems' (Hanson, 1982: 283). Information processing technologies are at the storm centre of all defence/war activities. In this light it ought not to be surprising to note that, from the development of radar systems, through the origination of the transistor, to a vast array of more recent Artificial Intelligence projects, it has been the demands of military agencies that have fuelled the growth of computer communications and information handling technologies (Braun and MacDonald, 1978).

Professor John Erikson articulates a central issue:

Modern military operations are not to do with weapons. . . . They are to do with information, command, control. Information does things. It fires weapons. It tells them where to go. The signals network is the key thing. If you want to disarm the world . . . get rid of all the computers.

It's not about the muscle, the strong arm of the warrior. It is his nervous system that matters. Signals and Communications.

(quoted in the *Guardian*, 11 November, 1982: 4)

Modern war/defence is quintessentially about information. Whether it is a matter of building 'intelligence' into 'smart weapons' such as Cruise missiles, whether it is developing ground-to-air weapons capable of 'locking' on to the target, whether the concern is to introduce electronic measures, counter-measures, and counter-counter-measures: whether it is any or all of these things *information is at the core of modern military affairs.*

The electronic weapons that are developed of course incorporate information/intelligence into the technologies themselves. For example, Exocet missiles, so devastatingly put to use in the Falklands War, contain such sophisticated electronics that the user can 'fire and forget': their speed, range, accuracy and skimming capacities make them practically 'invulnerable' by enemy defences.

The importance of systematic and routine surveillance and communications networks to any credible defence/war operation deserves to be emphasised. Supporting and integrating electronic weapons systems are increasingly indispensable command, control, communications and information (C3I) networks. These, bolstered by a mind-boggling collection of satellite technologies and more flexible support systems such as airborne warning and control systems (AWACS), are crucial to maintain what have been referred to as peacetime and wartime 'information regimes' (Bracken, 1983). In non-war conditions they are used to manage routine military affairs and logistics, to facilitate long and short range communications, and to observe and analyse the actions of enemies and potential enemies. In war conditions they are all the more essential. With speed of detection and response at a premium in modern war because of the rapidity with which any attack may be executed, armed forces are spending billions of dollars to improve the detection and communications stages of their defence networks. Only the most advanced computers could offer any hope of achieving this, and accordingly computer systems are massive investments. Such is the urgency for speed of response that defence interests are now seeking ways to computerise the decision-making processes of military encounters, for example by developing 'launch on warning' programs with help from expert systems, and perhaps even totally automated systems. 'There is no technological reason,' writes Frank Barnaby (1986: 2), 'why warfare should not become completely automated, fought with machines and computerized missiles with no direct human intervention.'

As such a route is followed, the military becomes reliant on information systems to conduct its affairs: without such communications, control and surveillance structures the military would be disabled even before engaging in battle. Precisely such reliance has led military planners to conceive a strategy of 'decapitation': if one side aims to be victorious in a nuclear confrontation, then it is useless trying to attack missile silos or ground troops since they are so large and dispersed that a retaliatory strike would be inevitable. Instead, it must pre-emptively destroy the enemy's C3I systems, hoping to eliminate disconnected forces in the ensuing turmoil (Ford, 1985: 122–146).

All this, of course, directly involves surveillance, since much of the C3I networks are premised on the gathering and analysis of a very wide variety of information. Here let us underline the most obvious target of surveillance, the monitoring of enemies and potential enemies, which emanates from the obligation of all nation states to be prepared to defend their territories against real or perceived threat.

There may be some who, pointing to the end of the Cold War, believe that the imperatives that drive defence/war institutions have been removed. Against this, it is crucial to realise that 'the preconditions for intelligence as a permanent government function lie in the modern state system' (Whitaker, 1992: 121). Because it is the first duty of any government to protect its frontiers, there is an insatiable hunger for information about anything affecting national interests and sight of the Communist monster is not essential to stimulate this appetite. The consequence has been the construction of a massive system of interlinked technologies to routinely and continuously monitor and inspect events and activities – military and civilian – around the globe (Richelson and Ball, 1986).

Satellites are a linchpin of surveillance activities. According to two commentators, it is getting to the point where the military 'would be struck deaf, dumb and blind should their satellites be destroyed' (Jasani and Lee, 1984: 4). Illustrative of the importance of these C3I systems are the accounts of the reasons for and consequences of the destruction of Korean Airlines flight 007 and its 269 passengers on 1 September 1983. Several writers argue that the flight was intended to enter Soviet territory to get the Russians to activate new defence systems and to open lines of communication between Moscow and their Far East nuclear command stations so that orbiting American satellites could spy on them (Johnson, 1986). Even those who argue that the plane's trespass was due to navigational error do not dissent from the fact that the United States enjoyed an intelligence bonanza when the Soviet C3I system was alerted (Hersch, 1986).

Necessarily, these systems are hidden from public view, secrecy being essential to ensure security from the enemy. Thus is constructed an anonymous and unexaminable, national and world-wide, web of surveillance and transmission of messages between defence agencies. As William E. Burrows (1986: 22) observes:

> the system that does all this watching and listening is so pervasively secret –
> so black – that no individual . . . knows all of its hidden parts, the products
> they collect, or the real extent of the widely dispersed and deeply buried
> budget that keeps the entire operation functioning.

The surveillance machine is not only directed against external enemies. Given the nation state's susceptibility to internal assault there is a powerful impulse towards searching out 'subversives'. In pursuit of the enemy within, Britain's security service, MI5, has at its Mayfair headquarters a Joint Computer Bureau with the capacity to hold 20 million records, files on one million people, and a network of 200 terminals accessing the mainframe (Campbell and Connor, 1986: 274). Leaks and occasional exposés have revealed that surveillance is exercised on trade unionists, numerous Labour MPs, CND activists, educationalists and media personnel, as well on what might be thought to be more obvious candidates (cf. Leigh, 1980; Massiter, 1985; Hollingsworth and Norton-Taylor, 1988). In addition, MI5 works in association with the Special Branch of the police force, thereby extending its information-gathering network nationwide (Davies and Black, 1984: 19). The security services also have access on request to a vast array of data banks, including the Police National Computer, Inland Revenue records, British Telecom files, and data held by the Department of Health and Social Security.

In sum, what we witness is a powerful force impelling the growth of surveillance systems which emanates from the nation state's duty to safeguard its frontiers. In a world divided by national frontiers there is, unavoidably, a built-in pressure towards the construction of effective defence/war machines. And because nations are often in conflict and, even more common, in situations of potential conflict with one another, what 'effective' means is always subject to change. However, what remains constant is the impulse to garner, adapt and act upon the best possible information about real and putative enemies within and without. On one level this is manifest in the design and development of increasingly sophisticated 'smart' weapons technologies. On another, it results in the establishment of communications networks of scarcely comprehensible complexity. And on still another, though this intersects with both of these, there is a sustained pressure towards observation and tracking of anything that may be conceived as a danger to national security.

Through the work of Tony Giddens we get a clearer idea of the contribution of war and the nation state to the development of surveillance. Thereby we can better appreciate the limitations of classical social theory's emphasis on either 'capitalism' or 'industrialism' as an account of the emergence of modernity. However, Giddens does not suggest that capitalist priorities play no part in the spread of surveillance. Here it is relevant to take note of a recent article by Stansfield Turner (1991), one-time Director of the CIA (1977–1981), now concerned with finding a role for the agency in a post Communist world.

Mr Turner has no hesitation in urging that the CIA continues its role of surveilling other nations. Indeed, while acknowledging the diminished Soviet threat, he is adamant that it must still be closely scrutinised (not least because it is an unstable region, in possession of nuclear weapons, which could threaten US interests), and he goes on to identify plenty of other dangers, especially in the Third World, to US security, which must therefore be closely monitored (p.152).

What is most interesting about Mr Turner's article is his wish to extend the remit of the CIA from military and political surveillance to economic spying. We must, he insists, 'redefine "national security" by assigning economic strength greater prominence', because 'the pre-eminent threat to US national security now lies in the economic sphere' (p.151). What economic intelligence constitutes 'can range from the broad trends that foreign businesses are pursuing, all the way to what individual foreign competitors are bidding against US corporations on specific contracts overseas' (p.152). With such a broad remit, and a recommendation from so highly placed a man as Stansfield Turner for 'a more symbiotic relationship between the worlds of intelligence and business' (ibid.), clearly we ought not to overestimate the degree of autonomy of the nation state's defence and its commercial concerns.

POLYARCHY, CITIZENSHIP AND SURVEILLANCE

There is another way in which the nation state has impelled the expansion of surveillance, one that has links with military enterprise, but which carries fewer of the chilling associations. This is the concern of the state with its citizens, notably how people have come to attain citizenship rights and duties, and how these are delivered and enforced. Integral to the development of these rights and duties has been the spread, in the nation state, of polyarchy, broadly understood as the extension of democratic means of governance.

To understand this better one needs to return to the foundation of the nation state. Forged in warfare, often of an internecine and drawn-out kind, a priority of any sovereign power which intends to rule a given territory is what Giddens calls 'internal pacification'. Bluntly, order and stability must be achieved within one's borders as a prerequisite of securing one's external frontiers. No doubt, in the early days, 'internal pacification' could take the form of compulsion by force of arms, but much more than this is required of a state which has ambitions for long-term survival. Minimally, the state must know its subjects not least because it may well require some of them to be conscripted to fight off foreign attackers. Further, each nation state needs knowledge of its subjects so that it may effectively administer taxation. And both of these needs mean that some form of census is a requisite of all nation states – hence surveillance is a priority from the outset. To this extent the Domesday exercise, undertaken in the late eleventh century, is a prime example of an early form of surveillance: the Conqueror, eager to consolidate and gain from his military incursion, immediately instigated a system aimed at getting to know the populace – their identities, addresses and, of course, their wealth.

Since then, though of course with a chequered history, it is possible to trace the extension of ways of monitoring the internal population. During the late nineteenth and early twentieth centuries especially there was an extraordinary expansion of official statistics, meticulously gathered by increasingly sophisticated techniques, ranging from regular census materials to figures on anything from educational performance to employment patterns in particular areas of the country (cf. Hacking, 1990). Undeniably, the information thus collected is fascinating as a means of comprehending the changing character of British society, but it is also, and crucially, a requirement of the nation state which must take responsibility for matters such as taxation, usually determines educational provision, and may even have a regional economic strategy.

This may be to jump ahead of the argument. A resonant theme of the development of the nation state is, as we have seen, the need to defend militarily its borders and, to this end, a census, however rudimentary, is essential since the state must be able to levy taxes and to call upon its male subjects to withstand invaders and even to take part in expansionist gambits. But something else is required. In order to get young men to fight on a state's behalf a good deal more than knowledge of their abode and occupations is necessary. The nation state must offer them something more tangible.

Giddens, drawing on the pioneering ideas of T. H. Marshall (1973) suggests that this may be conceived as a form of 'contract' between the nation state and its members. In essence the proposal is that, in return for fighting for the nation, subjects have achieved a variety of *citizenship rights:* for example the right, as a citizen, to the protection of the state from attack by outside forces, or the right to carry a passport which allows free entry into one's host nation and support at one's embassies in foreign countries.

Several sociologists have observed the links between the waging of war and the coming of social reforms which have expanded citizenship rights. It seems that, when many men are called upon to make large sacrifices in war, and, still more tellingly, when the entire population must engage in a 'total war', then there is a powerful impetus to social reform. In Britain, for instance, it is fairly easy to discern a link between the mobilisation for and sacrifices made during the First World War and the promise of a 'land fit for heroes' which brought the vote to all men over the age of twenty-one, improved the rights of women, and introduced modest state pensions. Similarly, it is clear that there is a connection between the circumstances endured by soldiers and citizens between 1939 and 1945 and the collection of reforms that followed and established the Welfare State. Of course, these are not precise causal relations, and there are many other contributory factors, but what does appear to be indisputable is that there is a correlation between mobilisation for and sacrifice in war and social reform.

Today, as citizens, people living in the nation state have many rights and duties. For instance, they have a right to education, to vote, to hold a passport, to a minimum level of income, to health treatment and so on. They also have duties, as citizens, to pay taxes which are levied and, if called upon, to fight for the country.

The key point is quite simple: out of the 'contract' between the nation state and its members has emerged a battery of citizenship rights and duties. The main connection with surveillance concerns how these are to be delivered and collected. The nation state, under whose umbrella citizenship operates, must develop administrative means to meet these additional responsibilities. And it is this, broadly speaking the growth of the modern social democratic state, which is an especially powerful force for surveillance.

It is so because the administration of citizenship rights and duties requires the meticulous individuation of the state's members. Electoral registers require the development of data bases recording age and residence of the entire population; social services need detailed records of people's circumstances, from housing conditions and medical histories to information about their dependants; the Inland Revenue creates gigantic files which detail the economic circumstances of everyone in the country; throughout one's school years records are constructed describing attainments, developments, continuities and changes; programmes to mitigate the worst consequences of poverty require a great deal of information on those unfortunate enough to be considered eligible. . . . As Paddy Hillyard and Janie Percy-Smith (1988) put it: 'The delivery of welfare benefits and services is at the heart of the system of mass surveillance, because it is here that the processes of classification, information gathering and recording are constantly multiplying' (p.172).

DANGERS OF SURVEILLANCE

The nation state's propensity towards surveillance, propelled either or by security needs or the rights and duties of its citizens, or by both has generated a host of questions and problems. To the fore have been the concerns of civil libertarians who express considerable apprehension about the advance of surveillance. There is an extensive literature highlighting problems such as the creation of police files on people which may be misused in the vetting of juries or which may even lead to wrongful arrest (e.g. Manwaring-White, 1983; Marx, 1988).

Of particular and pressing concern are two related issues. One is the fear that agencies may have access to files collected for other purposes, for instance when police forces may gain access to employment, medical or banking records. The other concerns the more general issue of melding disparate data bases. With the computerisation of most state (and very many other) surveillance files comes the possibility of linking once-separate information. While there are considerable inhibitions placed in the way of making these connections, the potential is obvious of an 'electronic identity card', based perhaps on one's National Insurance Number, capable of constructing a 'total portrait' of particular individuals. Were agencies able to access say, medical, educational, tax, employment, banking and criminal records, it is clear that an individual profile of considerable complexity and detail could be constructed. This process will be facilitated by the decisions of government departments to introduce compatible computer networks, made in the mid-1980s and actively pursued since. Such a development, inescapably attractive to government officials seeking efficiency and/or better control, massively escalates the surveillance already undertaken.

From this one may be drawn to conceiving of modernity by way of the metaphor of the *panopticon*. This notion was taken up by Foucault (1979) from the original ideas of Jeremy Bentham on the design of prisons, hospitals and asylums. For those unfamiliar with the metaphor, the panopticon refers to an architectural design by Bentham whereby custodians, located in a central (and usually darkened) position, could observe prisoners or patients each of whom inhabited a separate, usually illuminated, cell positioned on the circumference. This design is adopted by Foucault as a metaphor for modern life, one which suggests that surveillance allows the construction of an panopticon *without* physical walls. Nowadays, courtesy of modern electronics technologies, people are watched, but they neither communicate with others who are observed nor can they see who it is who is doing the surveillance. The idea is that surveillance by unseen observers is an integral feature of advanced societies.

It is perhaps easy to exaggerate, and it would be a mistake to suggest that those who are surveilled in the 'disciplinary society' (Foucault) have no contact with other subjects of surveillance. A good deal of the information gathered about citizens from centralised sources such as the census does feed back to people and, indeed, enables them to reflexively monitor their own position, prospects and lifestyles (Giddens, 1991). For example, information on earnings levels, crime rates, or divorce patterns is useful not only to state officials, but also to individuals searching to make sense of, and to establish perhaps new directions in, their own lives.

It is important not to jettison the notion of the panopticon because it insistently reminds us of the overweening ambition of the state to see *everything*, and of the ways in which power and the accumulation of information are intimately connected. Manuel De Landa (1991), reflecting on military surveillance, refers to its 'machine vision' (p.204) manifested in things like telecommunications interceptions and satellite observation of foreign terrains, where the surveillance is automatic. Programmes are established which trawl all communications within a defined category, or satellites monitor anything and everything that falls under their 'footprint'. De Landa describes the sophisticated software that is developed to allow machines to decipher satellite photographs which pick up virtually everything beneath them, as well as the systems created to facilitate analysis of bugged communications. Looking at all such trends, he is drawn to describe it as a 'Panspectron', something 'one may call the new non-optical intelligence-acquisition machine' (p.205).

Elsewhere Benedict Anderson (1991), detailing ways in which the colonial state combined map and census 'to put space under the same surveillance which the census-makers were trying to impose on persons' (p.173), evokes the metaphor of the 'Glass House' to comprehend the colonialists' ambition 'of total surveyability' (p.184) of everything they had conquered and aimed to rule. A form of speech quite as evocative as Bentham's panopticon, the 'Glass House' conceptualises 'a human landscape of perfect visibility; the condition of [which] was that everyone, every-thing, had (as it were) a serial number' (pp.184–185).

In considering the insatiable desire of the nation state to surveille everything, how can we ignore Mr Stansfield Turner of the CIA? Advocating investment in 'a robust network of satellites with a variety of sensors', he argues that 'it would be a bargain because the ability to peer anywhere, anytime, is bound to be of great value in the uncertain new world ahead' (1991: 151). He urges continued and extended surveillance from the skies of obvious targets such as the former USSR and volatile Third World countries, but lengthens his list of targets to include 'world-wide checks on weapons and a country's intentions for using them [which] will require a highly intrusive monitoring and verification system'. Not only this, 'national rivalries' must also be watched – wherever they might appear – since 'the United States can make good choices only by being well informed'. Turner adds that 'there are growing global problems such as terrorism, ecological abuse and drug trafficking that will require the United States to collect a diversity of informa-tion' (p.153).

As if this brief were somehow insufficient, Turner recommends that his system of satellites should garner economic intelligence (as well as political and military) about friendly nations. He finds it a trifle embarrassing to be urging the US to spy on nations such as Britain and Japan, but finds relief in the fact that satellites and discreet communications intercepts (seeing but unseen) can undertake the surveil-lance without the allies noticing:

> as we increase emphasis on securing economic intelligence, we will have to spy on the more developed countries – our allies and friends with whom we compete economically – but to whom we turn first for political and military assistance in a crisis. This means that . . . the United States will want to look to those impersonal technical systems, primarily satellite photography and intercepts.

> (ibid.: 154)

With this we are close indeed to Bentham's panopticon.

These prospects may be chilling, but they are certainly not imaginings from the wild side of science fiction. They are logical extensions of the imperative to surveille which lies at the heart of the nation state (Gandy, 1993). This surveillance is an integral feature of all modern societies and 'there is no obvious and simple political programme to develop in coping with [it]' (Giddens, 1985: 310). Further-more, given the organisational requirements and their accompanying observational forms which provide so much of our way of life, we have to conclude, also with Giddens, that 'aspects of totalitarian rule are a threat' in all advanced societies precisely because surveillance is 'maximized in the modern state' (p.310).

CORPORATE SURVEILLANCE

In drawing on Tony Giddens' work to lead us towards a clearer understanding of state surveillance, we should not forget capitalist enterprises' contribution to the trend. Giddens himself does not ignore the part played by capitalist endeavour, stating tartly that 'surveillance in the capitalist enterprise is the key to management' (1987: 175). We have come across Stansfield Turner's advocacy of a 'more symbiotic relationship between the worlds of intelligence and business' (1991: 152), which is a reminder of the potential unity of nation states and capitalist activity. And the conservative Edward Shils does not exempt capitalist corporations from his list of trespassers on privacy. Indeed, it is Shils' (1975: 335) observation that corporations have been amongst the main intruders into personal lives that alerts us to the role of modern management in the development of surveillance.

A good case can be made for the view that management, fundamentally a phenomenon only of the twentieth century, is in essence a category of information work, a central purpose of which is to surveille exhaustively the corporation's spheres of action, the better to plan and operationalise strategies which ensure capital's best return on investment. As the pivotal figure of Scientific Management, F. W. Taylor, put it: the *raison d'être* of managers is to act as information specialists – ideally as monopolists – as close observers, analysts and planners of capital's interests. Taylor voiced the ambition, if not the achievement, of management:

> the deliberate gathering in on the part of those on management's side of all the great mass of traditional knowledge. . . .The duty of gathering in of all this great mass of traditional knowledge and recording it, tabulating it, and, in many cases, finally reducing it to laws, and even to mathematical formulae, is voluntarily assumed by the scientific managers.
>
> (Taylor, 1947: 40)

A key starting point for management, and the particular concern of Taylor, was the production process, long a problem, but becoming particularly intractable with the development of large plants and workforces in the late nineteenth and early twentieth centuries. There is a very extensive literature on the response of corporate capitalism to this, focusing on the growth of Scientific Management, which emphasises that managers were designated to perform the 'brainwork' (Taylor) of organisations, the better to exert effective control over what they manage. It is not my purpose here to judge the relative success or failure of specific management strategies: clearly, these are variable. However, it is important to note that surveillance of the production process has grown extensively since the early days of personal supervision through to recent innovations such as Computer Numerical Control (CNC), robotics and, of course, computer terminals where start times, production rates, presence at work stations, quality controls, and the like can be individuated and recorded automatically (Shaiken, 1985).

Modern management monitors production very closely as a requisite of much else, but the purview of management nowadays is necessarily much wider than work processes (Fox, 1989). Central to understanding this is realisation that corporate capitalism has expanded this century in three key ways. First, corporations have grown spatially, so that typically the leading corporations have at least a national, and usually a transnational, presence. Second, corporations have consolidated into fewer and much bigger players: typically a cluster of organisations dominate major market segments. Third, and easily overlooked, corporations have burrowed deeper into the fabric of society, both by developing the outlet networks which are readily seen in most towns, and by replacing much self and neighbourly provision with purchasable goods and services.

One major consequence of these trends, which amount to what has been called the 'incorporation of society' (Trachtenberg, 1982), is that they pose challenges for managers which, in order to be met effectively, rely upon sound intelligence being gathered. In short, surveillance of much more than the shop floor is nowadays a requirement of effective corporate activity. There are many dimensions of this, ranging from monitoring of currency fluctuations to political circumstances in host nations, but here I would centre on the development of surveillance of customers and, indeed, the wider public.

The expansion of market research, both within and without corporations, is an index of management's need to know its clientele. Its methods of accessing the public are variable, including survey and interview materials, commissioned public opinion trawls, and careful pre-testing of goods prior to launch and, indeed, during their design and development. They are getting increasingly sophisticated as market researchers endeavour to find out more about the the 'lifestyles' of potential and actual customers (Martin, 1992).

A close cousin of such surveillance is credit checking agencies which, as well as enquiring about the financial standing of customers, often generate address lists of possible buyers for their corporate clients. The area within which these latter operate is somewhat clouded, but most readers will have received unsolicited post from companies which have bought their addresses from another organisation. The reasoning is simple: if a golf club has a membership list, then this information is very useful to corporations which specialise in golfing holidays or, more broadly, in sports clothing.

It is important to take cognisance of the heightening of this surveillance which has accompanied the spread of versatile electronic technologies. David Burnham (1983) alerted us to the phenomenon of 'transactional information' some years ago, and it is one with special pertinence for contemporary surveillance. This is 'a new category of information that automatically documents the daily lives of almost every person' (p.51) as they pick up the phone, cash a cheque, use a credit card, buy some groceries, hire a car, or even switch on a cable television set. Transactional information is that which is recorded routinely in the course of everyday activities. It is constructed with scarcely a second thought (and frequently automatically, at the flick of a switch, the dialling of a telephone number, as a secondary consequence of an action such as taking money from a bank's electronic teller).

When this ordinary, everyday information comes to be aggregated, it gives corporations quite detailed pictures of clients' lifestyles.

There is, of course, a sinister side to all of this, but here I want to stress the practical use of such surveillance to modern corporations. The transactional information that is amassed at Marks and Spencer whenever someone makes a purchase at their computerised tills tells the company, precisely, what is selling, how rapidly or slowly, in which locations – essential information to the managers of the organisation. Moreover, when the customer uses a Marks and Spencer company credit card, the information is that much richer because it contributes towards an individuated portrait of that person's spending habits, clothing and food tastes, even preferred shopping locations. It is a form of surveillance that can very helpfully enhance the company's marketing strategies – for example, advertising material can be judiciously targeted to particular types of customer. Much the same sort of case can be made for the increasingly popular cable television channels: a profile can be generated of programme preferences, regularity of watching, duration of viewing and so on which will be of considerable use to corporate clients.

CONCLUSION

In this chapter I have tried, drawing especially on the work of Tony Giddens, to delineate major contours of the spread of surveillance which may be said to account for much of the 'informatisation' of society. It is in the extension of the nation state and its intimate concerns with war and defence, in the growth of citizenship rights and duties, and in the extension of corporate capitalism (to which I return in the following chapter) that we can see what may be better termed, not the 'information', but the 'surveillance society'.

Perhaps to repeat unnecessarily, this is *not* to paint an Orwellian scenario, though it does contain strong echoes of 'Big Brother' in various dimensions, since it can be seen as a necessary corollary of the observational imperatives that accompany a more organised way of life.

Nonetheless, we ought not to ignore power in this account. Not everyone has the same capacity to influence the agencies of the state and the corporate structures which undertake surveillance. To fully appreciate the significance of surveillance we really must identify those who instigate and act upon such monitoring because, as Christopher Dandeker (1990) observes, power here is generally a matter of *power over subjects*. Such a concern usefully reminds us that a 'surveillance society' is at the same time also a 'disciplinary society'. On that note, we may turn, in the following chapter, to the school of thought which pays most attention to identifying the groups in society which exercise power *over* others.

5

INFORMATION AND ADVANCED CAPITALISM: HERBERT SCHILLER

Today there are many more images than ever before and, of course, a range of new media technologies transmitting them. Information networks now traverse the globe, operating in real time and handling volumes of information with an unprecedented velocity, which makes the telegram and telephony of the 1970s appear way out of date. Similarly, it is impossible to ignore the routine use of word processors and computerised work stations in offices, to be blind to the expansion of advertising and its metamorphosis into forms such as sports sponsorship, direct mail and corporate image promotion. In short, the 'information explosion' is a striking feature of contemporary life and any social analyst who ignores it risks not being taken seriously.

As we have seen, there are thinkers who believe this is indicative of a new 'information society' emerging. For such people novelty and change are keynotes to be struck and announced as decisive breaks with the past. Against these interpretations, I want to focus on Marxist (perhaps more appropriately Marxian)[1] analyses, centring on one thinker, Herbert Schiller, who acknowledges the increased importance of information in the current era, but also stresses its centrality to *ongoing* developments, arguing that information and communications are foundational elements of established and familiar capitalist endeavour.

Given the widespread opinion that Marxists hold to an outdated creed, it may seem odd to encounter a Marxian thinker who concedes, even stresses, that we are living in an era in which 'the production and dissemination of . . . "information" become major and indispensable activities, by any measure, in the overall system' (Schiller, 1976: 3). Such thinkers do insist on the resonance of well-tuned themes, but they are deeply aware of trends in the information domain. Along with Herbert Schiller, scholars such as Peter Golding, Graham Murdock and Nicholas Garnham in Britain, and Vincent Mosco and Stuart Ewen in the United States, offer a systematic and coherent analysis of advanced capitalism's reliance on and promotion of information and information technologies. These Marxist-informed accounts achieve more than enough credibility to merit serious attention.

Herbert I. Schiller (born 1919) is the most prominent figure amongst a group of Critical Theorists (something of a euphemism for Marxist-oriented scholarship in North America) commenting on trends in the information domain. He is a New-

York-raised intellectual who was radicalised by the slump of the inter-war years and by his experiences as a soldier in post-Second War Europe (Schiller, 1991b). Though he came rather late to academe, publishing his first book in 1969 and beginning to teach in the information field only a couple of years earlier, he has had a marked effect on perceptions of the 'information age'. This has come about not least from his attendance at conferences around the world, but it also stems from a regular output of books and articles (the most important of which are *Who Knows?* 1981, *Information and the Crisis Economy*, 1984, and *Culture Inc.* 1989a).

POLITICAL ECONOMY

Schiller was trained as an economist, though he has been a Professor of Communications at the University of California San Diego since 1970. This background and interest, combined with his own radical dispositions, is reflected in his central role in developing what has come to be known as the 'political economy' approach to communications and information issues. This has a number of key characteristics (cf. Golding and Murdock, 1991), three of which seem to me to be of special significance.

First, there is an insistence on looking behind information, say in the form of newspaper stories or television scripts, to the *structural* features which lie behind these media messages. Typically these are economic characteristics such as patterns of ownership, sources of advertising revenue, and audiences' spending capacities. In the view of political economists these structural elements profoundly constrain, say, the content of television news or the type of computer programs that are created. Second, 'political economy' approaches argue for a *systemic* analysis of information/communications. That is, they are at pains to locate particular phenomena, say a cable TV station or a software company, within the context of the functioning of an entire socio-economic system. Accordingly, political economists start from the operation of the *capitalist system* to assess the significance and likely trajectory of developments in the information realm. Another way of putting this is to say that the approach stresses the importance of *holistic* analysis, but – to pre-empt critics charging that this is a crude and closed approach where, since everything operates in ways subordinate to the overall 'system' nothing much can change – a third major feature comes to the fore. This is the emphasis on *history*, on the periodisation of trends and developments.

Schiller's starting point is that, in the current epoch of capitalism, information and communication have a pronounced significance in relation to the stability and health of the economic system. Echoing Hans Magnus Enzensberger, Schiller and like-minded thinkers regard 'the mind industry' as in many ways 'the key industry of the twentieth century' (Enzensberger, 1976: 10). Thus Herbert Schiller:

> There is no doubt that more information is being generated now than ever before. There is no doubt also that the machinery to generate this information, to store, retrieve, process and disseminate it, is of a quality and character

never before available. The actual infrastructure of information creating, storage and dissemination is remarkable.

(Schiller, 1983a)

Of course, this is also a starting point of other commentators, most of whom see it as the signal for a new sort of society. Schiller, however, will have none of this. With all the additional information and its virtuoso technologies, capitalism's priorities and pressures remain the same:

contrary to the notion that capitalism has been transcended, long prevailing imperatives of a market economy remain as determining as ever in the transformations occurring in the technological and informational spheres.

(Schiller, 1981: xii)

It is crucial to appreciate this emphasis of Marxian analysis: yes, there have been changes, many of them awesome, but capitalism and its concerns remain constant and primary. For instance, Douglas Kellner (1989b) acknowledges that 'there have been fundamental, dramatic changes in contemporary capitalism' (p.171). He favours the term 'techno-capitalism' as a description of the period when 'new technologies, electronics and computerization came to displace machines and mechanization, while information and knowledge came to play increasingly important roles in the production process, the organization of society and everyday life' (p.180). However, these novel developments neither outdate the central concepts of Critical Theory nor do they displace established capitalist priorities. Indeed, continues Kellner, the system remains fundamentally intact, so the terms used by an earlier generation of Marxist scholars (class, capital, commodification, profit etc.) are still salient.

An integral element of Marxian concern with the consequences for the information domain is the role of *power, control* and *interest*. Over twenty years ago Herbert Schiller insisted that the 'central questions concerning the character of, and prospects for, the new information technology are our familiar criteria: *for whose benefit and under whose control will it be implemented?'* (Schiller, 1973: 175. original emphasis). These remain central concerns for like-minded scholars, and characteristically they highlight issues which recurrently return us to established circumstances to explain the novel and, as we shall see, to emphasise the continuities of relationships that new technologies support. Typically Schillerish questions are: who initiates, develops and applies innovative information technologies? What opportunities do particular people have – and not have – to access and apply them? For what reasons and with what interests are changes advocated? To what end and with what consequences for others is the information domain expanding?

KEY ELEMENTS OF ARGUMENT

In the writing of Herbert Schiller there are at least three arguments that are given special emphasis. The first draws attention to the pertinence of *market criteria* in informational developments. In this view it is essential to recognise that information and communications innovations are decisively influenced by the market pressures of buying, selling and trading in order to make *profit*. To Professor Schiller (and also to his wife, Anita, a librarian who researches informational trends) the centrality of market principles is a powerful impulse towards the *commodification* of information, which means that it is, increasingly, made available only on condition that it is saleable. In this respect it is being treated like most other things in a capitalist society: 'Information today is being treated as a commodity. It is something which, like toothpaste, breakfast cereals and automobiles, is increasingly bought and sold' (Schiller and Schiller, 1982: 461).

The second argument insists that *class inequalities* are a major factor in the distribution, access to and capacity to generate information. Bluntly, class shapes who gets what information and what kind of information they may get. Thereby, depending on one's location in the stratification hierarchy, one may be a beneficiary or a loser in the 'information revolution'.

The third key contention of Herbert Schiller is that the society which is undergoing such momentous changes in the information and communications areas is one of *corporate capitalism* (Williams, 1961). That is, contemporary capitalism is dominated by corporate institutions, which are concentrated, chiefly oligopolistic, organisations that command a national and generally international reach. Wherever one cares to look, corporations dominate the scene with a few hundred commanding the heights of the economy (Barnet and Müller, 1975). For this reason, in Herbert Schiller's view, corporate capitalism's priorities are especially telling in the informational realm. At the top of its list of priorities is the principle that information and IT will be developed for *private* rather than for public ends.

Clearly these are established features of capitalism. Market criteria and class inequalities have been important elements of capitalism since its early days, and even corporate capitalism has a history extending well over a century (cf. Chandler, 1977), though many of its most distinctive forms have appeared in the late twentieth century. But to Herbert Schiller this is precisely the point: the capitalist system's long-established features are the key architectural elements of the so-called 'information society'. From this perspective those who consider that informational trends signify a break with the past are incredible since, asks Schiller, how can one expect the very forces that have generated information and IT to be superseded by what they have created?

The 'information society' therefore reflects capitalist imperatives – i.e. corporate and class concerns and market priorities are the decisive influences on the new computer communications facilities – and, simultaneously, these informational developments sustain and support capitalism. In this way Schiller accounts for the importance of information and IT in ways which at once identify how the history of capitalist development has affected the informational domain and, at the same time, how information has become an essential foundation of that historical development.

TRANSNATIONAL EMPIRE

We may get a better idea of how Professor Schiller sees things if we take time to review his views on the development of capitalism in the twentieth century. He is particularly alert to the fact that as corporate capitalism has grown in size and scope so too has it created what might be called a *transnational empire*. The automobile industry is today a global activity in which the likes of Ford, General Motors and Nissan are prominent; computers mean IBM and a cluster of smaller (but still huge) companies like Digital Equipment, Sperry-Univac and Apple; telecommunications means AT&T, ITT and similarly positioned and privileged giants. . . .

Information and its enabling technologies have been promoted by, and are essential to sustain, these developments in several respects. One stems from the fact that corporations which range the globe in pursuit of their business require a sophisticated computer communications infrastructure for their daily activities. It is unthinkable that a company with headquarters in New York could co-ordinate and control activities in perhaps fifty or sixty other countries (as well as diverse sites inside the United States) without a reliable and sophisticated information network.

To Schiller this indicates ways in which information is subordinated to corporate needs, but I suppose that a less committed observer might argue that the 'IT revolution' took place and just happened to suit corporate concerns, though over the years there has come about a corporate dependence on information networks. However, there are two objections to this line of reasoning. The first is that the information flowing within and between sites is of a particular kind, one which overwhelmingly expresses corporate priorities. The second comes from Dan Schiller (1982), when he argues that the genesis of the computer communications network – its locations, technical standards, pricing practices, access policies – characteristically has prioritised business over public interest criteria. Dan Schiller's account of the history of information networks reveals that corporate concerns have shaped its evolution, have channelled it in *these* directions rather than in *those*, while establishing it as a focal point of capitalist operations. Information was thus developed to suit corporate interests, though in the process corporations have become reliant on information flows.

Another way in which the information arena has been developed to further the goals and interests of transnational capitalist enterprise, while it has in turn become essential to sustain capitalism's health, is as a mechanism for selling. Herbert Schiller attests that the vast bulk of media imagery produced is made available only on market terms and is simultaneously intended to assist in the marketing of, primarily, American products. Thus the television productions, Hollywood movies, satellite broadcasting – the entertainment industry *tout court*, in which the USA plays the leading part – is organised on a commercial basis and functions to facilitate the marketing of goods and services. On the one hand, this is manifested in the construction of TV channels only where there is a viable commercial opportunity, and in the supply of television programming on the basis of commercial criteria –

most commonly a sufficiency of advertising revenue. This leaves its impress on content, resulting in a preponderance of sensationalist and action-packed adventures, soaps and serialisations, sports and more sports, intellectually undemanding and politically unthreatening programming, all of which is intended to command the largest possible audience ratings that most appeal to advertisers and corporate sponsors.

On the other hand, the global marketing of, say, Levi Jeans, Coca-Cola drinks, Ford cars or Marlboro cigarettes would be hard to imagine without the informational support of the mass media system (Janus, 1984). As far as Herbert Schiller is concerned this is of the deepest consequence (Mattelart, 1991). It is the starting point of any serious understanding that American media should be expected to laud the capitalist way of life – hence the beautiful homes depicted in so many programmes, the plethora of celebrities, the enviable lifestyles and opportunities. . . . To be sure, some popular programming does suggest a seamier side to contemporary America, notably the underbelly of the inner cities, but such programmes still retain a glamour and excitement which demonstrates something profoundly admirable to watchers in Seoul, Manilla or Seville. A primary aim of US media is not to educate the Indonesian, Italian or Indian in the mysteries of *Dallas, Bonanza* or *thirtysomething*; rather it is 'to open up markets and to get as large a chunk of the world market as possible' (Schiller, 1992: 1).

However, Herbert Schiller also highlights some rather more direct ways in which mass media, overwhelmingly emanating from the USA, give ideological support to its transnational empire. One key way stems from the prominent position enjoyed by the United States in the production and distribution of news. Being the major source of news reporting, it is perhaps not surprising that American media reflect the concerns of the home nation. The upshot is that 'free enterprise', 'free trade' and 'private ownership' are phrases widely used and conditions frequently advocated in the news services. Similarly, 'economic health' and 'industrial success' are defined by the terms and conditions prevailing in the capitalist economy – thus 'competition', 'markets' and 'business confidence' are terms unproblematically adopted to depict what is presumed to be the normal and desirable condition.

More importantly perhaps, world events and trends are covered from a distinctively metropolitan – usually American – perspective. Nations are examined in the news only to the degree to which events there have some likely or at least possible consequence for the USA – unless a disaster is of such proportions that it commands the news by virtue of its drama. For example, late in 1993 Somalia was prominent because US troops were killed by local militia, and Haiti received attention because events there involved US personnel. Similarly, Middle East affairs receive coverage chiefly when there is a 'crisis' with major implications for the United States and its allies.

Connectedly, the bulk of international news – actually 90 per cent – published by the world's press comes from just four Western news agencies, two of which are American (United Press International and Associated Press), one British (Reuters) and the other French (Agence France Presse). These reflect their bases' concerns; UPI devotes over two-thirds of its coverage to the United States, and only 1.8 per cent to Africa (Pilger, 1991b: 10). With such an imbalance of coverage American (and the Western nations more generally) do not need to put out crude messages such as 'West is best', 'the American Way', or 'support capitalist enterprise' to be ideologically functional. It is enough that they provide an over-whelmingly Western viewpoint on events, an agenda of items which is thoroughly metropolitan in focus, with the rest of the world covered primarily as a location of 'trouble' (mainly when that has implications for the dominant nations) such as 'war', 'coup d'état', 'disaster', and 'drought'. Hitting the news of the world as 'problems', they readily come to be presented either as dismayingly unreliable and prone to dramatic acts of violence or as subjects to be pitied when hit by yet another cyclone, volcanic eruption or crop failure.

We ought not to forget the technological superiority the West also enjoys (in satellites, telecommunications, computers, etc.) which provides an insuperable advantage in supporting its perspectives (Schiller, 1981: 25–45). This combines with American primacy in the entire range of entertainment: the movies are American, the television is American, and so too is most of the music business. It is the Western capitalist societies that have the finance for the films, the resources for putting together a global marketing campaign, the capability to create, store and distribute hours of soap operas. The ideological messages in this area are frequently unclear, occasionally nuanced, and at times even contrary to the espoused aims of private capital. Nonetheless, what is surely hard to dispute is that, in the round, the messages of American entertainment, whether it be *Little House on the Prairie*, the *US Masters Golf* from Augusta, or *Miami Vice* are supportive of America's self-perception as a desirable, indeed enviable, society which other nations would do well to emulate.

This is the perception of Herbert Schiller, who has been one of the most determined advocates of a *New World Information Order* (NWIO). From the premiss that, underlying the media representations, lie unequal structural relationships which divide the world's populations, Schiller's position logically follows. He insists that a requisite of giving voice to the poorer nations' struggles to improve their lot is to challenge 'information imperialism'. At the moment the world's information environment overwhelmingly emanates from the Western nations, especially America (McPhail, 1987). News, movies, music, education, and book publishing is pretty much a 'one-way street' (Varis, 1985; Nordenstreng and Varis, 1974). Even non-radical analysts accept that there is a 'media dependency' (Smith, 1980) on the West, and there are also a good many non-Marxian thinkers who are concerned about this situation and its possible consequences. In France, for in-stance, there is a long tradition which protests about the threat to cultural integrity from a preponderance of American-made media produce (Servan-Schreiber, 1968).

To Herbert Schiller all this constitutes 'cultural imperialism', an informational means of sustaining Western dominance (Tomlinson, 1991). He advocates a challenge to this – hence the call for a 'new world information order' which has had a marked effect in UNESCO (Nordenstreng, 1984) and which led to the US's withdrawal from that organisation when it leaned towards support for such a policy (Preston *et al.*, 1989).

It should be emphasised that we are not simply identifying here a pressure from without which bears down on the information domain. Quite the contrary: the maturation of corporate capitalism has been a process of which the information industry has been an integral and active part. Hence the history of the spread of corporate capitalism has also been a history of the spread of media corporations. And, just like corporate capitalism as a whole, media corporations have expanded in size, concentrated in numbers, frequently diversified their interests and moved decisively onto an international stage.

Two leading British 'political economy' scholars, Peter Golding and Graham Murdock, have assiduously chronicled these trends over two decades (Murdock and Golding, 1974; 1977a,1977b). The conclusion of their detailed empirical investigations of the shifts and strains in the British information industry is blunt: 'massive communications conglomerates' have been brought into being, 'with an unrivalled capacity to shape the symbolic environment which we all inhabit' (Murdock, 1990: 2). They do distinguish different types of media conglomerate, but all share common characteristics, most important of which is that they are gigantic capitalist corporations with national and international reach (Murdock, 1982).

Given such traits, reasons Herbert Schiller, we ought not to be surprised that contemporary mass media are enthusiastic supporters of the capitalist system. Indeed, given their operational principles and organisational forms, surely the odd thing is to account for those few areas of modern communications which do not whole-heartedly espouse capitalist values (cf. Dreier, 1982).

MARKET CRITERIA

Schiller's view is that the contemporary information environment is expressive of the interests and priorities of corporate capitalism and an essential component in sustaining the internationalist capitalist economy. However, there is a good deal more to the Marxian approach to information than this. We will be better able to appreciate the contribution if we exemplify ways in which central capitalist concerns make their influence felt on the 'informatisation' of society.

It is useful to begin with that key concern of capitalism – the market. Schiller's claim is that market principles, most emphatically the search for profit maximisation, are quite as telling in the informational realm as they are throughout capitalist society. As a rule, information will therefore be produced and made available only where there is the prospect of its being sold at a profit, and it will be produced most copiously and/or with greatest quality where the best opportunities for gain are evident. It follows that market pressures are decisive when it comes to determining what sort of information is to be produced, for whom, and on what conditions.

Those who contend that the market is the decisive force in capitalist societies insist that the products which become available themselves bear the impress of market values. A startling example of this was provided by the then Chairman of Thorn-EMI, a major British IT supplier, when he announced that his company's 'decision to withdraw from medical electronics was [because] there appeared little likelihood of achieving profits in the foreseeable future' (Thorn-EMI, 1980). In this instance the operative value was that Thorn-EMI perceived its interests to be best served by following a strategy whereby it concentrated on consumer entertainment products. Medical electronics were not felt to be supportive of the search for maximum profitability whereas television, video and other leisure products were.

The corporations which dominate the information industry operate unabashedly on market principles, and to this end they tailor their production to those areas which hold out the prospects of greatest reward. This point must confound those who believe that in the 'information age' information technologies are aloof from social influence at least in terms of their hardware (after all, goes the refrain, a PC can be used either to write sermons or hard pornography; in itself it is neither good nor bad since it is above social value), and that more information is intrinsically a good thing (it appears to be a deep-seated presumption that in and of itself more information is beneficial).

It must be disconcerting that this Critical Theory maxim looks behind the finished products that reach the market and asks: what were the priorities of the corporate suppliers at the research and development (R&D) stages? R&D budgets, nowadays multibillion-dollar annual commitments from players such as IBM, AT&T and Siemens, are committed to creating the next generation of technologies, but they are not given an open commitment by their paymasters. British Telecom (BT), for instance, spent some $400 million on R&D in 1992, but this was a carefully targeted investment. Two *Financial Times* journalists, observing that 'the days of research for its own sake are over', explained that they are 'a luxury that a commercially-oriented, competitive BT cannot afford' (Bradshaw and Taylor, 1993).

While clearly it is an imprecise relationship (R&D cannot guarantee the production of particular innovations), it is incontestable that private corporations decide to invest in research projects overwhelmingly for commercial reasons. One example from many: BT told the Monopolies and Mergers Commission in 1986 that it needed the green light for its purchase of a Canadian telecommunications equipment supplier company (Mitel) to enhance its commercial prospects. The reasoning offered was that, in order to expand its market share, BT needed to 'widen its product range by introducing new "key" products on which to base a variety of office systems for sale at home and overseas'. The company went on to explain that the R&D initiative to produce this new range was consciously targeted towards particular markets. Indeed, the Commission recorded that 'BT explained that a close relationship between research, development, manufacture and distribution was essential to move such new products quickly from the conceptual stage to the market place. BT's strategy therefore required it to have an integrated R&D and manufacturing unit' (Monopolies and Mergers Commission, 1986: 42).

It is unavoidable, given these pressures, that those areas which have most market appeal are most favoured in deciding on R&D distribution. Equally, this must mean that the products and services themselves are influenced by the application of market principles. For instance, the primary market for telecommunications businesses is the corporate one – where the big spenders are located. British Telecom, for example, receives about one-third of its revenues from just 300 customers (Locksley, 1984: 36), with over 70 per cent of traffic coming from business customers (Newman, 1986: 29), and international telecommunications is far and away BT's most profitable wing. Perhaps predictably, then, this is where one finds the most exciting information products and services. For the corporate sector all manner of premium services are available: from international telecommunications networks, PABX systems, facsimile services, data and text processing, to video-conference facilities. At the same time, at least 20 per cent of households in the UK do not possess even the basic telephone. It is understandable, given these circumstances, that the major British competitor to BT, Mercury (itself a subsidiary of transnational Cable and Wireless), was established with a clear remit to concentrate on the lucrative business realm. To this end its services centred on the City of London and were restricted to the large urban areas where most customers are found.

One paradoxical consequence of this prioritisation has been noted by former editor of *Computing* magazine. It is his estimation that most 'new' technology is, in fact, characteristically 'old' in that it complements existing products that have already proven their marketability. In this way the computer industry, Sharpe argues, offers a 'public mask of progress and the private face of conservatism' (Sharpe, n.d.: 111). For example, it is striking that most informational products for the home are actually enhancements of the television set. Video equipment, cable, computer games and suchlike are all founded on what has been a remarkably successful commercial technology. Why offer anything radically different when television has shown itself to be the public's favourite leisure technology?

Those readers who feel that such an outcome is an inevitability driven by an internal logic of technological innovation need to exercise some imagination here. There is really no technical reason either why home IT should be built around the television set or why programming should be so emphatically entertainment oriented. The most telling pressure has surely been that this was where and how the most lucrative sales would be made; accordingly, domestic IT/information was pushed and pulled in directions dictated by the market. Predictably, this results in familiar products and programming. As Sharpe comments:

> Alternative uses of technology are sought out by alternative groups. But they are few and far between. They mostly fail because the technology is not aimed at alternative uses, it is not developed to engender real change: for better or worse, it is developed to preserve.
>
> (Sharpe, n.d.: 4)

When one comes to examine more closely the actual information that has increased in such quantity in recent years one can easily enough fail to see the impress of market criteria. Since it is so popular to presume that more information is in itself advantageous, one rarely asks about the role of the market and some of the possibly negative consequences of this pressure. But it is useful to reflect critically on the dictum that all information is enlightening, in some way an advance on a – presumed to have been less 'informed', and thereby more ignorant – previous condition. Scepticism about the value of ever more television programming of an escapist kind readily springs to mind here and I suppose this is something about which many readers might concur. Less obvious, however, is to query one of the most remarked upon phenomena of the 'information explosion': the prodigious expansion of data bases available in real time within and across nations. No one has very precise figures, but what is known is that those data bases are aimed predominantly at a limited, if lucrative, market. Not surprisingly, perhaps, by the mid-1980s financial information, which represented over half of the electronic information sold in the United States, was 'the driving force of the industry' (Snoddy, 1986: 13). Typical database holdings are credit checks, price listings, energy reports, mergers and acquisitions, precious metals and world insurance. Not only is this information specialised; increasingly it is also targeted at particular market segments. So, for example, Lloyd's Maritime Data Network tracks the cargoes and movements of every commercial ocean-going vessel, compiling an awesome data base, though the information is directed at only a few hundred clients – shippers.

The designers of on-line information services have endeavoured to appeal to corporate clients since these have an identifiable need for real time business information and, tellingly, they have the ability to pay the premium rates that have fuelled the rapid rise of 'information factories' like TRW, Telerate, Quotron and Datastream (Schiller, 1981: 35).

On-line information services are at this time scarcely addressed to the wider public. In fact, only about one-third goes to the non-business consumer (*Business Week*, 25 August 1986: 52). Of course, subject to having the appropriate technology to hand, the ordinary citizen may use these data bases – provided he or she has the wherewithal to pay the average access costs of something like £100 per hour, plus the premium rates for particular services, and assuming, of course, that the specialised information stored appeals. On whatever count, market principles are likely to exclude all but a small minority of the public.

It is this which leads Professor Schiller to ask exasperatedly:

What kind of information today is being produced at incredible levels of sophistication? Stock market prices, commodity prices, currency information. You have big private data producers, all kinds of brokers . . . who have their video monitors and are plugged into information systems which give them incredible arrays of highly specific information, but this is all related to how you can make more money in the stock market . . . how you can shift

funds in and out of the country . . . that's where most of this information is
going and who is receiving it.

(Schiller, 1990b: 3)

David Dickson (1984) extends this argument in his history of science and technology – key knowledge realms – since the Second World War. He identifies two
elements, namely the corporate sector and the military, as the critical determinants
of innovation. To Herbert Schiller these are reducible to one, since it is his
conviction that the military's responsibility is to protect and preserve the capitalist
system and its market ethos. He writes that 'the military's preoccupation with
communication and computers and satellites . . . is not some generalized interest in
advanced technology. The mission of the USA's Armed Forces is to serve and
protect a world system of economic organization, directed by and of benefit to
powerful private aggregations of capital' (Schiller, 1984a: 382). The military might
make enormous demands on information, but since this is to bolster the capitalist
empire world-wide, then the fundamental shaper of the informational domain is the
market imperative at the heart of capitalist enterprise to which the military dedicates
itself. It is in this light that we can better appreciate Schiller's summary judgement
of the 'information society'. Far from being an evolutionary development of
intrinsic beneficence, it is expressive of capital's commitment to the commercial
ethic. Hence

> What is called the 'information society' is, in fact, the production, processing,
> and transmission of a very large amount of data about all sorts of matters –
> individual and national, social and commercial, economic and military. Most
> of the data are produced to meet very specific needs of super-corporations,
> national government bureaucracies, and the military establishments of the
> advanced industrial state.

(Schiller, 1981: 25)

Dickson extends this theme when he identifies three main phases of America's
science policy. The first, in the immediate post-war years, was dominated by the
priority of gearing scientific endeavour to the needs of military and nuclear power.
During the 1960s and 1970s there was a discernible switch, with social criteria
playing a more central role and health and environmental concerns making a
significant input to science policy. The third – and continuing – phase began in the
late 1970s and reveals an emphasis on meeting economic and military requirements.
By the early 1980s the guiding principle was decidedly 'the contribution of science
to the competitive strength of American industry and to military technology'
(Dickson, 1984: 17). This has resulted in science increasingly being regarded as
'an economic commodity' (ibid.: 33) and in the language of the boardroom and
corporate planning intruding into the heart of scientific activity. Today, attests
Dickson, innovation is guided by the principle that one will produce only that which
will contribute to profit. Hence routine reference is made to 'knowledge capital',
suggesting that scientists and technologists are regarded as factors of investment

85

from whom capital expects an appropriate return. From this perspective even scientists employed in academe come to be regarded as 'entrepreneurs' and are encouraged to co-operate closely with business people to create commercially viable products.

Dickson insists that this emphasis on the goal of success in the market necessarily directs scientific and technological knowledge away from alternative guiding goals such as public health, service to the local community, improving the quality of work experiences, or supporting the environment. The consequence is that universities, institutions at one time committed, at least in part, to wider community needs as well as the pursuit of knowledge for its own sake, have increasingly changed direction, dedicating themselves to research aimed at improving the commercial competitiveness of industry, thereby assuming that the marketplace is the appropriate arbiter of technological change.

Political programmes which have sought the *privatisation* of once-publicly owned utilities and the *deregulation* of one-time state-directed organisations have had a marked effect on the information domain. They have been openly trumpeted as the application of market practices by their advocates, as the most appropriate way to encourage efficiency and effectiveness (private ownership promising personal interest in resources and responsiveness to customers coming from this as well as from the primacy of buyers), and simultaneously as a means of introducing competition (and hence improved services) into previously monopolistic realms. Across Europe, the United States and the Far East, with variations resulting from local circumstances and histories, strategies for making the informational realm responsive to and dependent on market criteria have been put in place over the last decade or so (cf. Nguyen, 1985) with this twin element at their foundation.

Major effects have been evident, especially in telecommunications (Adonis, 1993). From the outset of its establishment in 1981 from the break-up of its state-owned parent the Post Office, British Telecom has operated on distinctively commercial lines, prioritising customers with the deepest purses (i.e. corporate and large government sectors) in its development of new and existing services and in taking measures aimed at ensuring its success as a capitalist enterprise.

In the days preceding its 1980s strategies, though its policy was rarely articulated, telecommunications in Britain operated with what may be called a loose 'public service' ethos. This guided the provision of services, aiming for universal geographical availability, non-discriminatory access, and a pricing policy that aspired towards 'reasonable costs or affordability' (OECD, 1991: 26), which was achieved by a complex system of cross-subsidy of discrete points on the network from lucrative urban and international links. The telecommunications monopoly also played an important role in supporting the British electronics industry by purchasing over 80 per cent of its equipment from these domestic sources, thereby acting to all intents and purposes as an arm of government economic strategy.

However, the market-oriented policies introduced during the Thatcher years encouraged deregulation and promptly took away the 'natural monopoly' of British

Telecom. In response, Mercury came into existence from private capital – with a mission, *not* to supply an alternative telephone service, but rather to win *business* traffic, easily telecommunications' major market. Since Mercury has little over 3 per cent of market share, its chief significance is not primarily as a competitor, but more as an indication of the new priorities prevailing in telecommunications.

BT's subsequent privatisation announced a renewed commercial emphasis in the organisation, one it marked with a decisive orientation towards the business market. This was expressed in various ways. First, responding to Mercury's attempt to cream off major corporate customers, BT reduced its prices in those areas. The company was quick to complain that it was 'making losses on local access' which it had once supported by charging over the odds to business users. This had not, of course, been a problem before, but by 1990 Mercury, free from the burden of offering a universal service, was attacking the corporate market, gaining almost 30 per cent of the national call revenue from customers with 100 or more lines. Now BT moaned that 'high usage customers [i.e. corporations] pay too much for their telephone services' while BT itself 'fails to make an adequate return from about 80 per cent of customers [i.e. domestic users]' (British Telecom, 1990). The consequence of such a diagnosis was predictable: though following privatisation and deregulation some government control has remained in the form of Oftel (Office of Telecommunications) which set a formula to restrict BT's price rises, this was only an average ceiling. In practice domestic users' costs rose ahead of those charged to businesses. In the early 1990s, Oftel's formula restricted BT price increases to the retail price index less 6.25 per cent (and in 1993 this was raised to 7.5 per cent). However, flexibility within the formula enabled BT to adjust prices in 1991 so that the average consumer telephone bill went up 5 per cent while the price of business calls fell with volume discount packages agreed and up to 20 per cent price reductions on international calls.

Second, BT, now a private corporation aiming to maximise profit, made moves to enter the global telecommunications market. It purchased manufacturing facilities in North America and became less interested in buying equipment from British suppliers. Further, in 1993 BT invested $4.3 billion (£2.8 billion) for a 20 per cent stake in MCI (Microwave Communications Inc.), the second-largest US long-distance telecommunications company. The motive behind this action was to advance a market-oriented strategy which recognised that the fastest growth area of the market was increasingly international, and that the really critical international market was that made up of corporate traffic. BT was clear-minded about this, recognising that 'the largest customers . . . are typically multinational companies with branches throughout the developed world' (British Telecom, 1990: 6). Accordingly, BT now has a 'highly-focussed strategy of supplying networks and network-based services to multinational companies' (British Telecom, 1993: 25). The stake in MCI was intended to enable BT to become a global leader in the provision of corporate network services. There is no comparable push to improve services to everyday domestic users.

BT feels no embarrassment about its prioritisation of the business market since it reasons this 'will be the source of the improvements in service and in techniques which will subsequently feed down to the residential market' (British Telecom, 1990: 6). This is, of course, the 'trickle down' theory of economics applied to the 'information revolution'. Following its logic, BT launched its global venture, Syncordia, which had only nine customers two years later (yet revenues of $200 million). A key element of the MCI deal is that Syncordia merges with the new BT/MCI joint venture company (which will have an estimated investment of $1 billion), making Syncordia that much more appealing to large corporations since they will have access to a 'one-stop shop' capable of merging and integrating their dispersed communications activities.

Third, throughout the 1990s BT has reduced its staffing while increasing its revenues: from a peak workforce of about 250,000 in 1989, it had dropped to almost £150,000 by the end of 1993, while income rose during the same period from £11 billion to £13.3 billion.

None of this should be read as a complaint against BT. Rather it should be seen as an exemplification of the primary role in developments in the information domain of market principles and priorities. Largely freed from former restrictions stemming from its days as a publicly owned monopoly, Britain's telecommunications giant now acts like any other private corporation. Its every aim is to succeed in the market and its services and practices are tailored to that end. If that means price rises over the odds for ordinary householders, labour lay-offs, and targeting of the most wealthy clients for new information services, then so be it. That is simply the logic of the market and the reasonable response of an entrepreneurial management.

Finally, we might draw particular attention to the *constraints* this market milieu imposes on participants such as BT. It is easy to believe that the adoption of market practices is largely a matter of choice for companies such as BT, but this is far from the case. Indeed, there are massive pressures disposing them towards certain policies. One overwhelming imperative is that the provision and servicing of information networks, while crucial for corporations in their everyday operations, is an intensively competitive market which impels players to act in given ways. As BT (1990) has noted, while a 'world wide telecommunications industrial structure can be expected' to emerge over the next decade or so, it will be one established and operated by 'perhaps [only] four or five large providers competing in the global market place at the cutting edge of the industry' (p.6). BT has ambitions to be amongst that elite, but there it will confront the likes of America's AT&T, a much bigger entity (revenues currently in excess of $60 billion per annum), and one equally determined to capture a large part of a global network market expected to grow to around $1,400 billion before the millennium (Bannister and Tran, 1993: 40). Pursuing this market, AT&T has shed over 140,000 of its workforce in recent months and, now freed from the government edict to restrict itself to domestic telecommunications inside the United States (America's version of deregulation), has moved to set up a global telecommunications network. All this for a reason equally obvious to BT: the readily perceived market opportunities in international business customers which have the biggest budgets and largest demand for sophisticated telecommunications services. Equally obvious, however, is the realisation that to fail in, or even to fail to enter, the global telecommunications market is an unthinkable option for the major suppliers. So they too are pressured into a race over which they have little control.

The primacy of market criteria in the information domain has had other consequences. An important effect has been that the promotion of the marketplace has led to a decrease in support for key information institutions which for long have been dependent on public finance. I discuss this in the next chapter, so here will summarise the theme. Institutions such as museums and art galleries, libraries, government statistical services, the BBC, and the education system itself have all encountered, in face of the 'information explosion', sustained real cuts in funding as a result of preference for market-oriented policies.

It has been government policy in Britain for over a decade that the most effective way to encourage the 'information revolution' is to make it into a *business* (ITAP, 1983). To this end public subsidies have been reduced and commercial values prioritised across a range of information institutions. For Herbert Schiller this represents an 'effort to extend the commercialization of information into every existing space of the social sphere' (Schiller, 1987: 25). Familiar stories of restrictions on library opening hours, shortages of funds to buy books, the charging for access to exhibitions formerly free to the public, above-inflation increases in prices for government information, closure of non-economically viable courses in colleges and so on are manifest results of this prioritisation of the market in hitherto protected realms.

According to Schiller all this represents 'the progressive impoverishment of social and public space' (Schiller, 1989b), with serious consequences for the generation and availability of information. In his view what we are witnessing is 'a silent struggle being waged between those who wish to appropriate the country's information resources for private gain and those who favor the fullest availability' – and in this struggle '(t)he latter have been in steady retreat' (Schiller, 1985c, p.708).

It is difficult to dissent from the view that, as public subsidy is replaced by private interests (or not replaced at all) which seek to develop information for the market, or, less dramatically, where public funds are so reduced that the institutions themselves are driven towards private sources of funds to remain viable, then there are major effects on what information is created and on what terms it is made available.

CLASS INEQUALITIES

The pivotal role of the market in the informational realm means that information and information technologies are created for and made available to those able to pay for them. This does not mean, of course, that they are totally exclusive. Clearly, virtually all members of society have some access to information products and services, television, radio and newspapers being obvious examples. Indeed, since the market is open to all consumers most of what is offered is, in principle, available to anyone – at least to anyone with the wherewithal to pay for them. However, the fact that the market is the allocative mechanism means that it is responsive to a society differentiated by income and wealth. In other words, class inequalities – broadly, the hierarchical divisions of society – exercise a central pull in the 'information age'.

What Vincent Mosco (1989) describes as a 'pay-per society' spotlights the *ability to pay* factor as a determinant force in the generation of and access to information: the higher one is in the class system the richer and more versatile will be the information to which one has access. As one descends the social scale so one gets information of an increasingly inferior kind.

Herbert Schiller (1983a) endorses this position, identifying as the 'chief executors' of the 'information revolution' – by virtue of their capabilities to afford the most expensive and leading-edge products of the IT /information industries – three institutions: the military machine, large private corporations and national governments. In this he finds support from business consultants Butler Cox and Partners (*Financial Times*, 11 April, 1983) who estimate that 80 per cent of the European IT market is accounted for by corporate and state outlets, with the 'general public' (i.e. everyone else apart from these two privileged groups) making up the remainder.

The centrality of ability-to-pay criteria, and the close linkage these have with class inequalities, leads Herbert Schiller to distinguish between the 'information rich' and the 'information poor', within and between nations (Schiller, 1983b: 88). This is easily enough illustrated. In countries such as Britain and the United States it is striking that, for the 'general public', the 'information revolution' means more television. Not only have the major developments been, in all essentials, enhancements of the TV monitor (cable, home computer, video etc.), they have also been programmed with a very familiar product – entertainment. The reasons for this are not hard to find. They lie in the fabulous success of television over the years (household saturation of equipment, a tremendous vehicle for advertising, entertainment shows relatively cheap to produce and very appealing). It must be remembered that here mass sales are essential since each household is, in relative terms, a poor source of revenue for the information industry. Those addressing the domestic realm must aim to supply a mass market since it is only when individual homes are aggregated as the 'general public' that they have any real market attraction. Once they are aggregated, however, the 'general public' must be offered information products which are undifferentiated – hence the familiar television monitor and the plethora of game and chat shows, movies and sport. Further, the 'general public' has proven itself reluctant to pay anything direct for TV programming – that has been subsidised by the advertiser and/or sponsor. Again though, with rare exceptions, advertisers who use television are interested in reaching mass audiences which in turn impels the programming towards 'more of the same' to ensure multimillion audiences. Any idea that the information needs of households may be variegated and sophisticated is lost, the major conduit for information provision being dedicated to entertainment, lowest-common-denominator programmes.

Much the same story pertains to cable and satellite services. While much has been written about the prospects of television responding effectively to the differential needs of the public, with thirty and more channels offering drama for those interested in theatre, ballet for those drawn to dance, news and current affairs for those keen on politics, and education for those wanting to improve themselves, the real history has shown, in the words of Bruce Springsteen, '57 channels and nothin' on'. Overwhelmingly, cable television channels offer entertainment programming: sport, soft pornography, action adventures, rock music video, and movies predominate. The fact is that the sophisticated and specialist channels dreamed about by the futurists in the early 1980s have come to naught, failing because they were too expensive for other than a tiny proportion of the population. The channels which have survived have tapped into the one rich vein – mass entertainment, where large audiences can be attracted for modest subscriptions or where advertising revenue can be commanded on promise of delivery of big numbers of viewers.

Comparable processes are visible between nations where differences of income lead to sharp information inequalities. The advanced nations where the world's wealth is concentrated are the major beneficiaries of the 'information revolution'. The poorer nations, wherein are located the majority of the world's population, are limited to the left-overs of the first world (for example, reruns of Hollywood serials), are dependent on what the affluent nations are willing to make available (for example what is produced from the news agencies), and may be further disadvantaged by the rich's monopoly of leading-edge information technologies such as satellites. These may monitor poorer nations from far above in the skies (for example for crop developments, mineralogical deposits, shoals of fish, even plain spying) and/or broadcast Western shows that undermine indigenous cultures and patterns of belief.

What is being suggested here is that, with the 'information revolution' being born into a class society, it is marked by existing inequalities and may indeed exacerbate them. Thus what has been called the 'information gap' may be widened, with those economically and educationally privileged able to extend their advantages by access to sophisticated information resources such as on-line data bases and advanced computer communications facilities, while those towards the bottom of the class system are increasingly swamped by what Schiller has termed 'garbage information' which diverts, amuses and gossips, but offers little information of value.

Here Schiller is observing that more information of itself does not necessarily enrich people's lives. On the contrary, the overriding determinant of information access and supply being ability to pay has meant that, for the majority, what is offered is cheap to produce, shallow, superficially appealing, mass information.

Surveying the surfeit of information offered in recent decades to the 'general public', from pulp fiction available now even in food stores, free newspapers delivered to every home, the explosive growth of 'junk mail', 24–hour-a-day television services, to the extension to every high street of video rental shops, the eminent journalist Carl Bernstein (1992) concluded that 'ordinary Americans are being stuffed with garbage'. Herbert Schiller vigorously concurs, arguing that 'we see and hear more and more about what is of less and less importance. The morning television "news", which provides an hour and a half of vacuous or irrelevant chatter, epitomizes the current situation' (Schiller,1987:30). In this sense the 'information revolution' has given the 'information poor' titillation about the collapse of royal marriages, daily opportunities to gawp at soap operas, graphic discussions of the sexual prowess of sportspeople – but precious little information that may let them in on the state of their society, the construction of other cultures, or the character of and reasons for their own situations.

CORPORATE CAPITALISM

In Professor Schiller's view the major beneficiary of the 'information revolution' is corporate capitalism. This has several crucial consequences for the information environment, each of which stems from its enormous wealth and central position in the modern economy. One is that information and allied technologies are developed and put in place with the corporate market uppermost in mind. The major computer installations, the front-end of telecommunications services, and the leading forms of electronic information processing are all to be found amongst corporations which have the ability to afford such things and, connectedly, have identifiable needs for ultra-sophisticated information facilities. For instance, as they have expanded in size, scale and space, it is clear that modern corporations have a built-in need for developed information networks and advanced systems of management control. Up-to-the-minute computerised technologies are a prerequisite of co-ordinating, of integrating and administering, organisations which typically have disparate locations.

It is truistic to say so, but still it needs to be said in face of so much celebration of the apparently supra-human origins of the new technologies: those who can pay for virtuoso IT seek out, and have provided for them, technologies which further their interests. The result of IT serving 'nicely the world business system's requirements' (Schiller, 1981: 16) is that it bolsters the powers of corporate capitalism within and without any particular society. And it does this in a wide variety of ways. For example, it enables companies to operate over distances using different workforces, responding to variable local circumstances (political, regional, economic etc.), with an efficacy unthinkable without real-time and sophisticated communications. Relatedly, it facilitates corporate strategies of 'decentralisation' of activities (i.e. slimming down corporate headquarters, and instructing subsidiary elements of the business to operate as 'independent' profit centres) while simultaneously bolstering centralised command because local sites can be easily observed, their performances tracked by a range of electronic techniques (e.g. precise sales records, records of productivity reaching down to individual employees).

Further, IT allows corporations to conduct their businesses globally with minimal concern for restrictions imposed by nation states. Corporations can operate telecommunications networks which offer them instantaneous economic transactions and real-time computer linkages along private lines which are removed from the scrutiny even of sovereign states. How, for instance, can a government, say, in Africa or India, know about the functioning of transnationals with bases in their country when information about the likes of Ford and IBM is passed between Detroit and Lagos or New York and Bombay in digital form through satellites owned by Western companies? There have long been questions asked about corporate practices such as 'transfer pricing' (i.e. internal accounting to ensure the best result for the corporation, whether or not, say, wage bills or investment commitments are a reflection of real costs in a given region); in an era of IT and associated electronic information flow it is almost impossible to conceive of getting accurate answers (cf. Murray, 1981).

Bubbling away amongst these observations on the power emanating from corporate access to information networks is another important ingredient – the spice that makes the 'information explosion' available only on *proprietary* grounds. One consequence, as we have seen, is that the corporate sector, with the most economic clout, is provided with the major information services. Another is that much information, once purchased, is then removed from the public view – or more likely never permitted to be seen – precisely because it is privately owned. Herbert Schiller thinks this is evident in contemporary America, where 'a great amount of information is withheld from the public because it is regarded and treated as proprietary by its corporate holders' (Schiller, 1991a: 44). Obvious examples of this principle – owners can do what they will with what they own – are information garnered by market research companies and research and development programmes undertaken by the corporate sector. Intellectual property, patenting and copyright are burgeoning areas of law in the 'information age': they are testament to the weight of proprietorial principles in this day and age.

Finally, it ought to be emphasised that corporate capital is not merely an external environment into which IT /information is being introduced. The 'information revolution' is not just being targeted at the corporate sector; it is also being managed and developed by corporate capital itself. In fact the information industry is amongst the most oligopolistic, gigantic and global of corporate businesses. A roll call of leading information companies announces some of the largest world corporations of the late twentieth century: IBM, Digital Equipment, ITT, Philips, Hitachi, Siemens, Sperry-Univac and General Electric are leading participants. Here a couple of examples are offered to illustrate something of the general pattern (Webster and Robins, 1986: 219–256).

Late in 1993 Bell Atlantic Corp., an American telephone company, tried to buy out Tele-Communications Inc. (TCI), the US's biggest cable operator, for an estimated \$33 billion. This 'most momentous deal of the decade in this decade of huge mergers, acquisitions and joint ventures' (Federal Communications Commissioner James Quello, quoted in *New York Herald Tribune*, 11 October, 1993: 11) represented the biggest takeover attempt in corporate history and would have resulted in a company ranking sixth on the Fortune 500 list of US companies (still behind information industry leaders IBM and General Electric). It proposed to bring together cable TV and telecommunications interests that already offer a range of services from entertainment to home shopping and enhanced communications. This deal floundered early in 1994 but, make no mistake, it is but the most dramatic of a series of mergers in this area which will continue. Associated with it was a multibillion-dollar TCI bid for Paramount Communications Inc. a major film company with an important backlist of film and video. The vision is to construct the 'information highways' of the next century, but the more immediate prospect is heightened control of corporate capital in their design and planning.

Earlier, in May 1991, AT&T, the world's biggest telecommunications company, announced that it was to merge with NCR, America's fifth largest computer outfit. The deal, valued in excess of $7.4 billion (*Financial Times*, 7 May, 1991), which marries computer and communications interests, is the most dramatic move into computing made by AT&T since it was released from government restrictions that confined it to telephony in the early 1980s. The rationale for the takeover was plainly voiced by AT&T Chairman Robert Allen: 'to meet what customers will need in the future – global computer networks as easy to use and as accessible as the telephone network today' (quoted in *Guardian*, 7 May, 1991). The meaning of the term 'Customers' here is presumed to be self-evident. But readers may usefully ponder to what extent their own informational needs require 'global information networks' and then, perhaps, they may wonder what particular customer need Mr Allen is preparing AT&T to meet.

CONSUMER CAPITALISM

The foregoing has concerned itself with showing how Professor Schiller and like-minded critics argue that the 'information society' is shaped by and most beneficial to advanced capitalism, to its market strictures, its structures of inequality and its corporate organisations. However, critics can go farther than this in two ways. The first, recently expanded by Oscar Gandy (1993), combines the theme of surveillance with an emphasis on the class and capitalist dimensions of the process. Thereby it is suggested that the informatisation of relationships is expressed by the increased monitoring of citizens in the interests of a distinct capitalist class. In these terms, for example, the state is a *capitalist* state, hence the spread of surveillance at its behest is a means of bolstering a subordinate class, by for instance building up files on active trade unionists, political subversives and radical thinkers, *en route* to more effectively restricting dissent. Similarly, the spread of surveillance for economic purposes is dedicated to strengthening the hold of capitalist relations (Mosco, 1989: 119–124). The second, connected, contention is that the 'information revolution' furthers capitalism by extending deeper into the everyday lives of people, hence encouraging the creation and consolidation of *consumer capitalism*.

This is taken to mean an individualistic (as opposed to collective) way of life, one in which people 'buy a life' (Lynd and Hanson, 1933) by paying personally for what they get. It entails a lifestyle which is home-centred to the detriment of civic relations, where people are predominantly passive (*consumers* of what capitalism has provided), and where hedonism and self-engrossment predominate and find encouragement. Consumer capitalism is thus an intensely private way of life, with public virtues such as neighbourliness, responsibility and social concern increasingly displaced by a concern for one's individual needs that it is felt, are most likely to be met by purchases in the store and shopping mall (and here, in the fantasy that in purchases we can find fulfilment of the self, is evidence of the collapse of the self itself: Lasch, 1985).

Informational developments are central to the spread of consumerism since they provide the means by which people are persuaded by corporate capitalism that it is both a desirable and an inevitable way of life. Through a sustained information barrage, attests Schiller, 'all spheres of human existence are subject to the intrusion of commercial values . . . the most important of which, clearly, is: CONSUME' (Schiller, 1992: 3). Here are some of the ways in which it is argued that consumer capitalism is encouraged by the 'information revolution'.

First, television is enhanced both to become a still more thorough means of *selling* goods and services to the individual buyer and to bolster the consumerist lifestyle. Television has already contributed much to the stay-at-home ethos of consumerism, and critics anticipate that flat-screen TVs, home entertainment systems, video and cable will spread this trend. Moreover, as these and other information technologies further penetrate the home, so too does their programming bear the imprint of those who would use it to further stimulate consumption. Advertisers and sponsors especially have created more, and more intensive, ways of getting across their messages to audiences: one thinks here of more careful targeting of images that can accompany subscription television, of the spread of advertorials, of judicious product placement amidst the television serial and movie.

Second, and related, the bulk of the programming itself, aside from the advertisements, serves to encourage a consumerist lifestyle. The symbols of success, beauty, fashion, popularity, approval and pleasure that are displayed in everyday television are presented to the public, which in response yearns for them and must seek for them on the market (Ewen and Ewen, 1982; Ewen, 1988).

These are, of course, arguments routinely presented in condemnation of the 'means of persuasion': the populace are brainwashed into chasing after 'false needs' that are manufactured to aid capitalism's perpetuation rather than in response to the wishes of ordinary people. The third argument, however, is less frequently made. This suggests that IT is exacerbating the tendency for the marketplace to replace self and communal organisation. Where once, for instance, people grew much of their own food in the garden, or perhaps made their own clothes, nowadays virtually all of our requirements are met at the supermarket or through the chain store

(Seabrook, 1982b). Similarly, it is suggested that television and TV-type technologies take away the responsibility of arranging one's own pleasures, replacing it with a new *dependency* on a machine which presents, in the main, diverting entertainment at which one gawks.

Fourth, new technologies allow greater surveillance of the wider public by corporations which are then in a better position to address messages of persuasion towards them. Years ago Dallas Smythe (1981) coined the term 'audience commodity' to draw attention to the way in which an important function of television was to deliver audiences to advertisers. The acid test for success was not to be found in the content of the programming, but in the numbers watching who could be sold to the advertiser. This continues today – and with a vengeance. For instance, free 'newspapers', delivered to every house in a given area, are not really intended to be a vehicle for informing householders of local news and events; their central concern is to be in a position to claim to deliver to the advertiser every house in a given neighbourhood. This is, of course, a pretty crude form of surveillance. Nonetheless, much more sophisticated forms come from the selling of data bases such as are held electronically by professional associations, clubs and sales records. Again, new technologies enable the ready development of profiles of customers and potential customers to be created by cross-referencing of such sources, to be followed by carefully addressed persuasion. Here subscription television has great possibilities since it will be able to segment viewers by channel, programme preferences and even by volume and regularity of watching. Examining this, Kevin Wilson (1988) coined the term 'cybernetic marketing' (p.43) to draw attention to the prospect of interactive technologies such as videotext being used for shopping from home via the television monitor. In such ways people will be ushered into still more privatised forms of life, while at the same time the suppliers will be able to construct, electronically, detailed portraits of every purchase. Thereby each transaction may be monitored, each programme watched recorded, contributing to a feedback loop that will result in more refined advertising and cognate material to further lock the audience into consumerism.

OBJECTIONS TO CRITICAL THEORY

There are a number of objections to be made to the Critical Theorist's position. One concerns the issue of policy. On the one hand, it is objected that it is hard to find in the writing of critics any practical propositions. On the other hand are those who proclaim that the collapse of Communist societies invalidates the critique. Since it is at least implicit in the writing of Schiller that a non-capitalist form of social organisation is possible – for instance he recurrently favours 'public information' over 'private' forms – and since the major experiments in collectivism have dramatically come to an end, then the Critical Theorists are asked to respond to this objection.

But the insights of Critical Theorists are neither obviated because they do not present an alternative policy nor are they nullified simply because non-capitalist regimes have fallen. The major value of the work of Schiller lies in its capacity to understand and explain the 'information age'. This is important not least because any alternative form of society that may be conceived must, if it is to be credible in any way, start with a sound grasp of the realities of the here and now. Very many future scenarios, and coming 'information society' sketches are commonplace, actually commence their analyses from idealistic premises such as the 'power and potential of technology' or, 'just imagine what we could do with all the information becoming available'. A distinct advantage of Schiller's accounts is that they remind us to start with an understanding of things as they are before we begin dreaming about alternatives.

Further, in explaining the genesis of the 'information age', Schiller's work presents the *possibility* of radically other ways of organising society. Seeing that the 'information society' has a *real human* history, that it is made by social forces, then by the same token we may imagine *another* way of making. To hold to the possibility of an alternative surely does not mean that one must endorse the only one – Communism – that has presented itself to date and subsequently failed.

Indeed, Herbert Schiller explicitly refuses this position. While he is depressed by the power and entrenched character of capitalism, he is also aware that

> those regimes of the world that have been organised according to non-market structures and arrangements since World War II are demonstrating a growing inability to provide alternative ways of producing information and cultural products. There is consequently, in my judgement, a dismaying acceptance of western media standards and models – as if these were genuine alternatives to what may well have been failed practices of their own.
>
> (Schiller, 1992: 4)

Another objection is that there is a strong sense of a 'fall from grace' in Marxian accounts. Demonstrating increased corporate influence, the spread of market relationships, and the development of consumerism, it is easy enough to conclude that things have got worse. The implication, for instance, is that a deluge of 'garbage information' has swamped what was once reliable knowledge, or that the spread of computer network facilities has led to more observation and thereby tighter control of workforces, citizens and individual consumers.

But we need to be sceptical of the notion of a 'decline', if only because we lack reliable historical and comparative knowledge. Certainly it may be shown that contemporary information is flawed in particular ways, but we must be careful not to assert that this necessarily makes it worse than hitherto. Further, as Anthony Giddens (1990, 1991) argues, the imposition of technologies for purposes of control or even to inflate the sales of corporate capital does not inevitably result in wholly negative consequences. It is possible that systems of surveillance *both* strengthen managerial control *and* increase choices for people.

On the subject of information inequalities, it may be noted that the radical critique works with a crude conception of the stratification system. To distinguish between the 'information rich' and the 'information poor' both avoids precise delineation of who these are and fails to consider the complexity and range of different positions in a class-divided society. In short, the model lacks sufficient sociological sophistication to allow consideration of gender, racial and ethnic differences, to say nothing of the expansion of non-manual groups and the resulting positions these occupy in the class hierarchy. Similarly, Schiller's attention to the corporate sector as the major beneficiary of the 'information revolution', while clearly being implicated in the class system, cannot be entirely accepted since institutional cannot be equated with personal wealth. That is, the 'information rich' as people are not synonymous with corporate capital, and the gap needs exploring in any acceptable analysis of information inequalities. Further, Schiller's under-developed conception of class fails to take account of cultural (as opposed to economic) capital, though in the realm of information/knowledge cultural capital such as higher education, access to libraries, and linguistic command may be decisive (compare say the affluent but ill-educated with the modestly rewarded but highly literate).

Another objection has to be the Critical Theorist's tendency to offer an 'all or nothing' view of information. Against this, it could be contended that, while there is a good deal of 'garbage information' in circulation, this does not necessarily mean that all the information directed at the general public is rubbish. Indeed, while the output of television may be seen to have expanded dramatically, and while the bulk of this may be a cocktail of chat, action adventures and soaps, in absolute terms it is possible to contend that high-quality information has also increased. In Britain, for instance, the introduction of Channel 4 in the early 1980s may have brought more American serials to the screen, but it has also increased the range and depth of television programming.

A cognate matter is the issue of the rapid take-up of the video cassette (VCR) recorder which in Britain at least has had an as yet immeasurable effect on viewing. One may speculate, however, that, where the major use of the VCR is for recording off air to watch on more convenient occasions ('time-switching'), this new tech-nology is allowing at least some audiences the flexibility to increase their access to high-quality information (arguably the sort scheduled for late-night minority audi-ences, put on too late for those who must rise before 8 a.m.). Much the same point may be made about pulp fiction. It is hard to look across the titles in W. H. Smiths and not feel a sense of dismay. Shallow and slick crime and soft pornography jostle for the big sales, readily making one yearn for Austen and Eliot. However, if the biggest sales are for pulp fiction, it is also the case that, in absolute terms at least, the classics are more available and more popular than ever thanks to the 'paperback revolution'.

Turning to information's alleged role in the spread of consumerism, it is as well to say at the outset that this is not a point restricted to Marxian critics. The identifica-tion of excessive individualism, the weakening of collective bonds, and the central role in this of market practices have been concerns of a wide range of thinkers covering a spectrum from Ortega y Gasset, T. S. Eliot and F. R. Leavis to Jeremy

Seabrook. A recurrent argument is that this requires manipulative information to instil in people 'false needs', to convince them that some personal weakness or hidden anxiety may be rectified by purchase of a given object such as shampoo or scent.

However, such positions have come under attack for several related reasons. At root there is some conception that once upon a time people had genuine needs which were met by simple things, that somehow life was more authentic, even if people were materially worse off. An image of 'plain living' but 'high thinking' is operative here, the idea of the working man coming home after a shift in the mine or factory to read his Cobbett or Hardy. And, of course, one objection is that life never was like that, that, for example, in the nineteenth-century fiction for the working man – when he read anything – was penny dreadful, sensationalised trivia about murder, rape, drink and fallen women (James, 1973).

Another objection refutes the presumption that people are duped by an avalanche of advertisements and related imagery. The belief of postmodernist (and other) adherents is that ordinary people are quite smart enough to see through the artificiality of consumerist images (they know holiday brochures don't tell the truth, that drinking beer doesn't guarantee friends and camaraderie), smart enough indeed to appreciate this imagery for the parodies it often offers, for its irony, its use of camera, colour or whatever.

Further, it may be a mistake to think only in terms of either privatised lifestyles or ones which are communally oriented. It is not inevitable that people who retreat into the home are thereby more self-engrossed, more cut off from neighbours and local affairs (Bellah *et al.*, 1985). Indeed, as Peter Saunders (1990) suggests, 'emphasis on the importance of home does not necessarily result in withdrawal from collective life outside the home, for it is possible for people to participate fully in both spheres of life' (p.283).

Finally, such a view suggests that imagery takes precedence over the products the advertisers are called upon to promote. But people do not buy Kit Kat chocolate because of the advertisements, but because they have an appealing taste. Similarly, it has to be said that a good many of the new information technologies are indeed superior products to their predecessors – for the domestic market one need think only of compact disc players, modern sound systems and even television sets which today are more attractive, provide better quality, and are more reliable than anything before. Moreover, it is surely also true that large numbers of people today buy consumer goods (from perfumes to entertainments), not because they have swallowed the puffery of the advertiser, but because they get genuine pleasure and increased self-esteem from these things.

CONCLUSION

It would be inappropriate to conclude on a negative note because I believe that there is a very great deal of value in Critical Theory, something surely evident from the bulk of this chapter. Several of its major emphases seem to me indispensable to an adequate understanding of the significance of information. Herbert Schiller's work especially, in starting with the real, substantive, world rather than with 'technological possibilities' or 'imagined futures', offers an important understanding of major dimensions of the role and significance of information and allied technologies.

The attention he draws to market criteria and corporate capitalism cannot but convince us of their pivotal role. Furthermore, he has a sharp eye for social inequalities which are *not* set to disappear in the 'information age'. Quite the contrary, he reveals, locally and globally, how these are key determinants of what kind of information is generated, in what circumstances, and to whose benefit. Finally, the identification of 'consumer capitalism', however much one might want to qualify the term and particular conditions, is a helpful reminder of just how much the informational realm is dedicated to the pursuit of *selling* to people who appear to be retreating further into privatised ways of life.

6

INFORMATION MANAGEMENT
AND MANIPULATION: JÜRGEN
HABERMAS AND THE DECLINE
OF THE PUBLIC SPHERE

There is a diverse group of commentators on the 'information society' who, while conceding that there is a lot more information in circulation nowadays, are rather unenthusiastic about announcements of the 'information age'. Such commentators tend to regard this information as being tainted, as having been interfered with by parties that have 'managed' its presentation, or that have 'packaged' it to 'persuade' people in favour of certain positions, or have 'manipulated' it to serve their own ends, or have produced it as a saleable commodity that is 'entertaining'. These thinkers lean towards the view that the 'information society' is one in which Saatchi and Saatchi's advertising campaigns, the Ministry of Defence's 'disinformation' strategies, Ford's public relations 'expert', the Parliamentary 'lobbyist', the judicious 'presenter' of government policy, and the 'official leak' from 'reliable sources' close to Downing Street all play a disproportionate role in the creation and dissemination of information. In its strongest versions, this interpretation suggests that the democratic process itself is undermined due to the inadequacies of the information made available to the public, since, if the citizenry is denied reliable information, then how can the ideal of a thoughtful, considered and knowledgeable electorate – a genuine democracy – be achieved?

THE PUBLIC SPHERE

In examining this critical approach I start with the work of the German social theorist Jürgen Habermas (born 1929) because his account of the *public sphere* has influenced much of this way of seeing. He developed the concept in one of his earliest books, though it was twenty-seven years before a translation of *The Structural Transformation of the Public Sphere: An Inquiry into a Category of Bourgeois Society* appeared in English in 1989. His argument is that, chiefly in eighteenth- and nineteenth-century Britain, the spread of capitalism allowed the emergence of a public sphere which subsequently entered a decline in the mid-to late twentieth century. It is taken to be *an arena, independent of government (even if in receipt of state funds) and also enjoying autonomy from partisan economic forces, which is dedicated to rational debate (i.e. to debate and discussion which is not 'interested', 'disguised' or 'manipulated') and which is both accessible to entry and open to inspection by the citizenry. It is here, in this public sphere, that public opinion is formed* (Holub, 1991: 2–8).

Information is at the core of this public sphere, the presumption being that within it actors make clear their positions in explicit argument and that their views are also made available to the wider public that it may have full access to the procedure. In perhaps its most elemental form, Parliamentary debate, and the publication of a verbatim record of Parliamentary proceedings in *Hansard*, expresses a central aspect of the public sphere, though clearly the role of communications media and other informational institutions such as libraries and government statistics can be seen to be important contributors to its effective functioning.

Readers will be able to conjure the ideal of the public sphere if they imagine open and honest Members of Parliament arguing cases in the chamber of the Commons, ably supported by dedicated civil servants who dispassionately amass relevant information about the subjects to be debated, with everything open to public inspection through a conscientious publications and press infrastructure prepared to report assiduously what goes on so that, come elections, the politicians may be called to account (and, indeed, that throughout terms of office public affairs may be transparent).

It will be useful to review Habermas' account of its history to understand more of its dynamics and direction. Habermas argues that the public sphere – or, more precisely, what he refers to as the 'bourgeois public sphere' – emerged due to key features of the expanding capitalist society in eighteenth-century Britain. Crucially, capitalist entrepreneurs were becoming affluent enough to struggle for and achieve independence from church and state. Formerly public life had been dominated by the clergy and the court where mannered display that celebrated feudal relations was the customary concern, but the growing wealth of capitalist achievers undermined this supremacy. In one way this occurred as they gave increased support to the world of 'letters' – theatre, art, coffee houses, novels and criticism – and thereby reduced dependence on patrons and stimulated the establishment of a sphere committed to critique which was separate from the traditional powers. As Habermas (1989) observes, here 'conversation [turned] into criticism and *bons mots* into arguments' (p.31).

From another direction came increased support for 'free speech' and Parliamentary reform as a consequence of market growth. As capitalism extended and consolidated, so did it gain greater independence from the state, and so too grew more calls for changes to the state, not least to widen representation so that policies could more effectively support the continuing expansion of the market economy. This struggle for Parliamentary reform, as a necessary corollary, was also a fight to increase the freedom of the press since it was axial to those who wished for reform that political life should be subject to greater public inspection. Significantly, *Hansard* was created in the mid-eighteenth century to provide an accurate record of proceedings in Parliament.

Alongside this was a protracted struggle to establish newspapers independent of the state, one much hindered by government antipathy, but facilitated by relatively cheap production costs. Revealingly, the press of the eighteenth and nineteenth centuries, while having a wide spread of opinion, was noticeably committed to very full coverage of Parliamentary matters, a sharp indication of the confluence of press and Parliamentary reform campaigns. Central to this mix of forces, of course, was the maturation of political opposition, something which stimulated the competition of argument and debate and which gelled with the pressure towards developing what Habermas terms 'rational-acceptable policies'.

The upshot of such developments was the formation of the 'bourgeois public sphere' by the mid nineteenth century with its characteristic features of open debate, critical scrutiny, full reportage, increased accessibility, and independence of actors from crude economic interest as well as from state control. Habermas emphasises that the fight for independence from the state was an essential constituent of the 'bourgeois public sphere'. That is, early capitalism was impelled to resist the established state – hence the centrality of struggles for a free press, for political reform, and for greater representation.

However, as the historical analysis proceeds, Habermas points to paradoxical features of the 'bourgeois public sphere' which led ultimately to what he calls its 'refeudalisation' in some areas. The first centres around the continuing *aggrandisement of capitalism*. While Habermas notes that there had long been a 'mutual infiltration' (p.141) of private property and the public sphere, his view is that a precarious balance was tilted towards the former during the closing decades of the nineteenth century. As capitalism grew in strength and influence, its enthusiasts moved from calls for reform of the established state towards a takeover of the state and use of it to further their own ends. In short, the *capitalist state* came into being: its adherents increasingly turned their backs on an agitational and argumentative role and used the state – now dominated by capital – to further their own ends. The result of the expansion of MPs' private directorships, of business financing of political parties and think tanks, and of the systematic lobbying of Parliament and public opinion by organised interests has been a reduction in the autonomy of the public sphere.

Habermas does not suggest that these trends represent a straightforward return to a previous epoch. His view is that, during the twentieth century especially, the spread of a public relations and lobbying culture is actually testament to the continuing salience of important elements of the public sphere, not least that it is acknowledgement of an area where political debate must be conducted to gain legitimacy. However, what public relations does, in entering public debate, is to disguise the interests it represents (cloaking them in appeals such as 'public welfare' and the 'national interest'), thus making contemporary debate a 'faked version' (Habermas, 1989: 195) of a genuine public sphere. It is in this sense that Habermas adopts the term 'refeudalisation', meaning to indicate ways in which public affairs become occasions for 'displays' of the powers that be (in a manner analogous to the medieval court) rather than spheres of contestation between different policies and outlooks.

A second, related, expression of 'refeudalisation' comes from changes within the *system of mass communications*. One needs to recollect that this is central to the effective operation of the public sphere since media allow scrutiny of, and thence widespread access to, public affairs. However, during this century the mass media have developed into monopoly capitalist organisations and, as they have done so, their key contribution as reliable disseminators of information about the public sphere is diminished. The media's function changes as they increasingly become arms of capitalist interest, shifting towards a role of public opinion former and away from that of information provider.

While these two features are expressive of the spread and strengthening of capitalism's hold over social relationships, there is something else which, from its early days in the eighteenth and nineteenth centuries, has fought to use the state to bolster the public sphere. It has frequently swum across the current that has swept us towards a mature capitalist economy. One thinks here of groups which have made an important contribution to the creation and spread of a *public service ethos* in modern society. Habermas observes that from its early days the 'bourgeois public sphere' has provided space for people who occupy a position between the market and government; between, that is, the economy and the polity. I refer here particularly to professionals such as academics, lawyers, doctors and some civil servants. It is arguable that, as capitalism consolidated its hold in the wider society and over the state itself, so did significant elements of these (and other) professions agitate, with some success, for state support to ensure that the public sphere was not overly damaged by capital's domination (Perkin, 1990: 359–404).

Habermas (1989) makes this point with broadcasting especially in mind, arguing that public broadcasting corporations were founded 'because otherwise their publicist function could not have been sufficiently protected from the encroachment of their capitalistic one' (p.188). But the argument that such were the tendencies towards takeover by capitalist interests that state involvement was required to guarantee the informational infrastructure for a viable public sphere can be extended to explain the character of several key institutions, notably public libraries, government statistical services, and museums and art galleries. Indeed, *the public service ethos, conceived as an outlook which, in the informational realm at least, was committed to dispassionate and neutral presentation of information and knowledge to the widest possible public, irrespective of people's abilities to pay, can be regarded as closely consonant with an orientation essential to the effective functioning of the public sphere.*

Reading Jürgen Habermas on the history of the public sphere, it becomes impossible to avoid the conclusion that its future is precarious. His account of its more recent development is gloomy: capitalism is victorious, the capacity for critical thought is minimal, there is no real space for a public sphere in an era of transnational media conglomerates and a pervasive culture of advertising. As far as information is concerned, communications corporations' overriding concern with the market means that their product is dedicated to the goal of generating maximum advertising revenue and supporting capitalist enterprise. As a result their content is chiefly lowest common denominator diversion: action adventure, trivi, sensationalism, personalisation of affairs, celebration of contemporary lifestyles. All this, appropriately hyped, appeals and sells, but its informational quality is negligible. What it does is no more (and no less) than subject its audiences 'to the soft compulsion of constant consumption training' (Habermas, 1989: 192).

Habermas goes still further: while the public sphere is weakened by the invasion of the advertising ethic, so too is it deeply wounded by the penetration of public relations. In this regard Habermas is especially sensitive to the career of Edward Bernays, the doyen of American 'opinion management', which he takes to be indicative of the demise of the public sphere. What Bernays and his many descendants represent is an end to the rational debate characteristic of the public sphere, this being subverted by the manipulative and disingenuous political operator. To Habermas this intrusion of PR marks the abandonment of the 'criteria of rationality' which once shaped public argument, such criteria being 'completely lacking in a consensus created by sophisticated opinion-molding' which reduces political life to 'showy pomp' before duped 'customers ready to follow' (p.195).

Habermas is unrelentingly glum. Universal suffrage may have brought each of us into the political realm, but it has also raised the primacy of opinion over the quality of reasoned argument. Worse than this weighing of the vote without assessing the validity of the issues, the extension to everyone of the suffrage coincided with the emergence of 'modern propaganda' (p.203), hence the capability to manage opinion in a 'manufactured public sphere' (p.217).

THE PUBLIC SPHERE AND INFORMATIONAL CHANGE

The idea of the public sphere offers an especially powerful and arresting vision of the role of information in a modern society (Curran, 1991: 33). From the premiss that public opinion is to be formed in an arena of open debate, it follows that the effectiveness of all this will be profoundly shaped by the quality, availability and communication of information. Bluntly, reliable and adequate information will facilitate sound discussion while poor information, still less tainted information, almost inevitably results in prejudicial decisions and inept debate. For this reason several commentators have drawn on the notion of the public sphere as a way of thinking about changes in the informational realm, using Habermas' concept as a means of evaluating what sort of information there has been in the past, how it has been transformed, and in what direction it may be moving.

More particularly, a conception – admittedly idealised – of the public sphere has been introduced into consideration of two connected and crucial areas of information. The first has been that of public service institutions such as the BBC and the library network, with writers concerned to argue that their informational function is being denuded especially, if not solely, by attempts to transform them into more market-oriented and organised operations. The second area is the wider context of contemporary communications, where commentators suggest that an increasing amount of unreliable and distorted information is being generated and conveyed.

Here the focus is on new systems of communication which stress commercial principles and end up purveying little but escapist entertainment, on the spread of interested information such as sponsorship, advertising and public relations, and on an increase in the use of information management by political parties, business corporations and other interest groups which inflates the role of propaganda in the contemporary information environment.

PUBLIC SERVICE INSTITUTIONS

Radio and television

Public service broadcasting organisations, manifested chiefly but by no means solely in the BBC (British Broadcasting Corporation), are amongst the most important informational institutions in the nation. The BBC, for instance, is at the heart of a great deal of political, cultural and social communication and, through television especially, is capable of reaching every member of the society.

Though it is rarely articulated (cf. Broadcasting Research Unit, 1986), public service broadcasting may be taken to be a type institutionally set apart from outside pressures of political, business and even audience demands in its day-to-day functioning, one not pressed by the imperatives of commercial operation, and one made available to, and produced for the benefit of, the community at large rather than those who either can afford to pay for subscription or who can attract advertisers and sponsorship revenue. It is committed to providing high quality and as comprehensive as possible services to the public, which is regarded as composed of diverse minorities which are to be catered for without endangering the provision of programming – news, current affairs, drama, documentary – aimed at the whole audience. Its practitioners are dedicated to providing services without disguising their motives and with a goal of enlightening audiences on a wide range of affairs and issues, from politics to domestic conduct. Of course, this is an ideal type definition, though the BBC, while it has interpreted public service with particular emphases over the years, has approximated to it. Several of these public service broadcasting characteristics echo Jürgen Habermas' depiction of the public sphere – notably perhaps the organisational location independent of both government and the market, the ethos of the public servants which stresses undistorted communication, and the service's availability to all, regardless of income or wealth.

The BBC was established to operate at a distance from commerce. This came about because of a peculiar unity of radicals and conservatives which allowed ready acceptance that the BBC be formed as a state institution aloof from the interests of private capital. Observers had witnessed the hucksterism, chaos and cacophony created by commitment in America to a free market in broadcasting, and their repugnance led in Britain to an odd domestic alliance: as historian A. J. P. Taylor (1965) noted, 'Conservatives liked authority; Labour disliked private enterprise' (p.233).

In this way the BBC was 'born in Britain as an instrument of parliament, as a kind of embassy of the national culture within the nation' (Smith, 1976: 54), granted a monopoly over broadcasting, and funded from an involuntary tax on wireless – later television – receivers (the licence fee). The formation of the BBC by Parliament and its aloofness from commerce had important consequences. It allowed for an emphasis, explicitly called for by the legislators, on broadcasting as a means of education as well as entertainment. Over the years this ethos – 'to inform, educate and entertain' – has been consolidated and expressed in much BBC output from news through to minority programmes of music, literature, drama and hobbies.

This cannot be translated straightforwardly into Habermas' terms of a public sphere dedicated to the furtherance of 'rational debate', but it has undeniably massively extended public awareness of issues and events beyond most people's experiences (and to this extent, whether reporting from overseas or depicting aspects of life in Britain long hidden from general view, it has performed an important democratising function). Paddy Scannell and David Cardiff (1991) argue that this extension of audiences' horizons involved a spread of 'reasonableness' in the sense that people were able, and called upon, to give reasons for what they did, how they lived, and what they believed. If these accounts were not necessarily 'rational' (since this term implies somehow a 'correct' account), they enriched public life in so far as they opened vistas at the same time as the BBC helped create a common culture in Britain amidst a diverse populace (Kumar, 1986).

The BBC, being a Parliamentary creation, has been profoundly affected in its practices and assumptions by the Parliamentary model. This has found expression in a presentation of political affairs that, on the whole, has limited itself to the boundaries of established party politics (the modulated 'balance' between Labour and Conservative parties) – with occasional adventures in drama and documentary – but at the least it aided the treatment of politics in a serious and considered manner. That is, public service broadcasting in Britain has always emphasised its role as an *informer* on public affairs. It has characteristically dedicated a great deal of time on the schedules to such coverage, in face of the appeal of presenting either cheaper or more popular programming.

The decisive influence of its founding Director-General Lord Reith, credibility achieved during the Second World War and its uncontested monopoly for some thirty years, were important factors in rooting the public service ethos in Britain (Briggs, 1985). There was the important additional factor that the BBC, notwithstanding attempts by governments to interfere (Tracey, 1978), has remained genuinely distanced from political dictates, being state-linked in contrast to state-directed systems where broadcasting has commonly been seen as an instrument of government policy. This has undoubtedly been essential to the sustenance amongst broadcasters of a commitment to political impartiality and to reporting as accurately and objectively as is possible.

Krishan Kumar (1977) has described the BBC's autonomy from commercial and political controls as 'holding the middle ground', a position which has certainly contributed to the 'quite unusual cultural importance that attaches to the BBC in Britain' (p.234) and that has attracted and been bolstered by the entry into broadcasting of many talented people instilled with a public service outlook and sceptical of the 'moving wallpaper' mode predominant in out-and-out commercial broadcasting systems (Burns, 1977).

In addition, the public service ethos of the BBC has had a marked influence on commercial broadcasting in Britain. Thus independent television, launched here in the mid-1950s following an intensive lobby, has from its outset had public service clauses injected into many of its activities. As James Curran and Jean Seaton (1988) observe, it 'was carefully modelled on the BBC [and] the traditions of public service were inherited by the new authority' (p.179). This is reflected in its Charter, which demands that it strive for impartiality in coverage, in the structure of its news services which are formally independent of the rest of its commercial activities, clauses in its contracts such as the requirement to show at least two-thirty minute current affairs programmes per week in peak time, and the financing of Channel 4 which puts it at arm's length from advertisers in order to protect its mission of reaching different audiences from previously established channels.

If broadcasting's public service roles set it to some degree apart from commercial imperatives, then it is important to say that this does not mean it has been aloof from outside pressures, able to operate, as it were, in the capacity of dispassionate and free-floating information provider. It could not do so since it is part of a society in which commerce is a powerful force, at the same time as the BBC is an institution created by the state and therefore susceptible to pressures that could be brought to bear by and on the state. Further, the recruitment of BBC personnel especially has come predominantly from a restricted social type (Oxbridge, arts graduates), a fact that has brought forward values and orientations that are scarcely representative of the diverse British public. Inevitably, broadcasting's evolution has been influenced by such pressures and constituents as these and the priorities they endeavour to establish.

However, this is not to say – as a good many left-and right-wing critics have alleged – that broadcasting is some sort of conduit for the powerful (the 'ruling class' for the Left, the quasi-aristocratic 'Establishment' for the Right). It has a distinctive autonomy from business and politics which has been constructed over the years, even though features of this independence have changed. In its early days under Reith the BBC was separate from government officials and disdainful of the business world, but it was an autocratically run organisation with an elitist orientation. Public service then was taken to mean the transmission of programmes that were considered worthy by custodians of what is now regarded as a rather outdated philosophy – in essence, Matthew Arnold's credo: 'the best that is known and thought in the world' (Reith, 1949: 116). In the 1960s circumstances were such as to allow public service to be interpreted in quite a daring manner while institutional independence was maintained. Under the directorship of Sir Hugh Greene (Tracey, 1983), at a time when the economy was booming, television ownership increasing and thereby ensuring the BBC an annual rise in revenue from additional licence fees, when the political climate was relatively tolerant and relaxed, public service was liable to be perceived as including challenging, innovative programming that could awake audiences to new and often disconcerting experiences.

Over time it is possible to trace changes in conceptions of public service broadcasting (Briggs, 1985), with an ethos of professionalism (public service broadcasting being seen as a matter of producing intelligent, well-made, unbiased, interesting and challenging programmes) coming to displace earlier emphases on paternal responsibility in the Reithian mode (Madge, 1989). However, while professional ethics are important to contemporary programme makers, they do not readily provide them with a public philosophy of broadcasting with which to respond to sharp attacks on the BBC. Furthermore, with hindsight we can see that public service broadcasting depended, in part at least, on the presumption of a unified – or potentially united – audience. For good or ill, since the late 1960s the divisions amongst audiences have become very evident and have made it difficult to speak without heavy qualification of a 'general public', giving rise to some hesitancy and indecision in broadcasting and leaving it more vulnerable to assault from hostile critics.

Since the late 1970s we have been experiencing what has been called a 'crisis of public service broadcasting'. It is a crisis which many perceive to be being resolved in a diminution of broadcasting's public sphere functions. There have been two major fronts on which this crisis has been fought, the political and the economic. On one side there have been vociferous attacks on broadcasters from those who regard them as a part of a 'new class' of privileged, smug and state-supported elites who are both 'leftists' and disposed towards 'nannying' the wider public, and yet 'accountable' neither to government, nor to private capital, nor even to the audiences whose licence fees keep the BBC going. On another side has emerged an economic critique which contends that the BBC is profligate with public funds, takes money without offering any recourse to those taxpayers who provide it, and which urges a new sovereignty to the 'consumer' who ought to be 'free to choose' what programming is to be provided (cf. Barnett and Curry, 1994).

These sides have combined in an assault which has led to sustained real reductions in the budgets of the BBC, a series of political interventions ranging from complaints about 'bias' about specific programmes (there are many instances, from alleged left-wing prejudice on Radio 4's *Today* news programme to recurrent complaints of shortcomings in coverage of Northern Irish affairs), to peremptory dismissal of a Director-General (Alasdair Milne) in 1987, to appointment of a conservative, combative and interventionist Chair of Governors (Marmaduke Hussey), to introduction of market mechanisms and commercial practices (the 'internal market', short-term contracts to many once-tenured staff, a fiercely directive managerialism inside the corporation : Tully, 1993), to close scrutiny from a government-instigated investigative committee, headed by a renowned free market economist, of the financing of the BBC (Peacock, 1986).

Behind all this, of course, is the enthusiasm for the market which has been so much a feature of recent times. The demise of public service broadcasting, therefore, is most often cast in terms of enthusiasm for 'competition' (liberalisation and deregulation) and 'privatisation' (ending state support in favour of private share-holding). It is of a part with this critique that the influential right-wing Adam Smith Institute (1993) proposed that the BBC should be privatised (cf. Hargreaves, 1993). As things turned out, the BBC achieved an extension of its licence fee to 2001 in the summer of 1994, which led to a bout of self-congratulation from 'modernising' Director-General John Birt who contended that the 'pain' [sic] of 25 per cent cuts in staffing, the introduction of 'producer choice', an orientation towards the 'global marketplace', and co-operative ventures with the Pearson media group, had saved the service from out-and-out commercialisation. It was dismay at such measures that surely led Dennis Potter (1994) to 'fear the time is near when we must save not the BBC from itself but public service broadcasting from the BBC' (p.54).

While the BBC is the major focus of attention amidst these changes, conse-quences for British commercial television ought not to be neglected. Independent Television in Britain was marked by the impress of public service demands, especially in strictures about the kind, quality and scheduling of news and current affairs programmes. These have often been placed, surely against the wishes of the more zealous television entrepreneurs, in peak time slots, the most significant of all being the nightly *News at Ten* broadcast at 10 p.m. The changed climate of British television was forecast by Mr Paul Jackson (1992), Director of Program-ming at Carlton Television. 'Given the commercial realities,' he warned, 'we won't have the latitude in future to find excuses for programmes that don't earn their keep. Programmes will not survive in the new ITV if they do not pay their way.' Unless news and current affairs can deliver audiences of a scale that appeals to advertisers (and they may, continued Mr Jackson, if they become 'more popular' and/or draw in 'quality', i.e. wealthy, viewers), then such informative programmes will be pushed to the sidelines of early evening or very late night (as in the USA) where they will not interrupt provenly popular television such as movies, soaps and game shows.

From yet another direction comes erosion of public service broadcasting insti-tutions by new means of delivery, notably satellite and cable television services, especially in Britain in the guise of Rupert Murdoch's B Sky B television service and its diet of 'entertainment' (sport, movies and 'family' programmes). The fear is that once the audience share of public service channels falls to 30 per cent or so, then its support from involuntary taxation and its claim to address the 'general public' become untenable.

If one seeks to discern the direction in which British broadcasting is moving, then one must surely look to the United States (Barnouw, 1978) because it is, in key respects, a model which guides the British government's information policies (though emulation does not extend to support for investigative journalism; in fact British disposition towards the 'strong state' – of which more below – has resulted in marked increases in censorship over the past decade and a half). In such a milieu, where the ratings largely determine media content, public service broadcasting must be hard pressed to survive.

The prospect is for more support for broadcasting coming from private funds, whether advertising, sponsorship or subscriptions, and for less from the public purse. With this transfer comes promotion of commercial criteria in programming, with the upshot that audience size and/or spending power (with occasional prestige projects backed by sponsors in search of reflected status) are the primary concerns. Content is unavoidably influenced by these emphases, with most often an increase in entertainment-centred shows as opposed to 'serious' and/or 'minority' concerns such as news and current affairs (though these are likely to be made more 'entertaining') and intellectually challenging drama.

To critics of this trend what we are seeing is an undermining of public service broadcasting and, with it, the weakening of its public sphere roles. While the prospect is for more 'wall-to-wall *Dallas*', for more emulation of American television's 'cultural wasteland', it is possible that some high-quality programming will be available via perhaps new forms of delivery or even by subscription. However, these will either be niche markets or they will be restricted to those groups with the wherewithal to afford requisite subscription fees, something which undermines the public sphere principle of information being available to everyone, irrespective of ability to pay.

A crucial issue today is whether the quality of information provided by broadcasting is declining and whether it is likely to continue to do so. For market enthusiasts 'narrowcasting' promises much more and much more accurately targeted information going to variegated and pluralistic customers. To thinkers influenced by Habermas, while there is no doubt that there is a much greater quantity of information generated on television and radio, it has not led – and it will not lead – to greater quality of information or to genuine choices for listeners and viewers. This is because the market generates trivia, or concentrates power in the hands of media moguls, or segments audiences by bank account so that quality information is limited to the better-off sections of society. Whichever way we look at it, the days of public service broadcasting and key public sphere features appear numbered.

Public libraries

The public library network is arguably the nearest thing we have in Britain to an achieved public sphere. There are well over 5,000 public libraries in the nation, reaching into pretty well every sizeable habitation.[1] The network features several of Habermas' public sphere elements, including the following:

- Information is made available to everyone, access being guaranteed as without cost to individuals. Membership is free to all who live, work or study in the local area, and public libraries must provide free books for loan, access to reference materials, and must have reasonable opening hours which facilitate access.
- The library service is publicly funded from taxation gathered centrally and locally, but its operation is independent of political interest, being instructed,

under the 1964 Public Libraries and Museums Act, 'to provide a comprehensive and efficient library service for all persons desiring to make use' of it. Should one's local library not hold the information for which one is searching, then the national system of inter-library loan, supported by the existence of designated copyright libraries, may satisfy one's requirements.

- The library network is staffed by professional librarians who provide expert advice to users as a public service, without prejudice against persons and without hidden motives (Library Association, 1983).

Several factors have contributed to the growth of public libraries, from upper-class philanthropy, paternalist sympathies, fear of the untutored masses and a desire to increase literacy rates, to a wish to open up educational opportunities by providing learning resources to the disadvantaged (Allred, 1978). Whatever divided these motives and aspirations, what lay behind them all was an important conception of information. That is, public libraries were formed and developed on the basis of a notion that information was a resource which belonged to everyone rather than a commodity which might be proprietary and hence privately owned. It followed that, since information and indeed knowledge could not be exclusively owned, it should be available freely to those who wished to gain access to it, a conception which appears to have been at the core of the establishment and operation of the public library system in Britain. It is fundamental to the public library network that if people want information they ought to have help in getting it and not be penalised in that search.

There is considerable evidence that public libraries are both popular and highly valued by the general public. Most people belong to a library and, in Britain, annually around 650 million loans are made and over a quarter of adults visit a public library every month (*Social Trends*, 1992, 10.12: 181). However, the public library system has come, noticeably over the last decade or so, under sustained challenge on both philosophical and practical grounds. Serious attacks have been made on the underlying premiss that information ought to be free to users of the library, and policies have been put in place that have pressured libraries increasingly to charge for their services.

The recent history of public libraries is one of cuts and commercialisation combined with a forceful critique of their public service aspirations. It is well known that, since the late 1970s, government policies have endeavoured to reduce public expenditure and to promote the application of market mechanisms as a means of providing services. As a result, libraries have had to cope with continued real reductions in their resources. Services have been curtailed: newspaper subscriptions have been cancelled, periodicals rigorously reduced, opening hours have shrunk and, in general, there are fewer volumes on the shelves (especially up-to-date books). Every library user will have noticed the depletion of newspapers and periodicals; it is calculated that there was a 30 per cent reduction in the number of libraries open for more than 45 hours per week between 1980 and 1991 (*Cultural Trends*, 1992 a, table 1.11: 14); and there is an escalating trend to sell off 'outdated' materials (and thus deplete the library's stock) to raise funds to finance the continuation of existing services (West, 1992).

Reductions in funds have pressured librarians to look to other sources of revenue. They have been encouraged along this route by a minister who advises librarians to 'look beyond their traditional sources of funds and consider whether some costs may be recovered from users, or whether private sponsorship, or even private investment in new services, is possible' *(Report by the Minister for the Arts on Library and Information Matters during 1983)*. With this come moves towards supplying information via the private sponsor and on its basis of its market appeal.

This new orientation was articulated in an influential report from the Adam Smith Institute, *Ex Libris*, published in 1986. In accord with the Institute's firm support for market allocation of services (Harris, 1978), this charged that 'the fundamental concept of providing libraries free at the public expense has seldom been challenged or its consequences examined' (Adam Smith Institute, 1986: 1). It lamented that free library provision had driven commercial lending libraries out of business and it urged the reintroduction of commercial criteria in information supply. To the forefront of its recommendations were charges for users and a move towards privatisation of the library world.

Underpinning these pressures towards marketisation has been a sharp critique of public libraries, one which comes now from the Right of the political spectrum, but which often draws on criticisms once made most vociferously by the Left. Perhaps most prominently, the free library service is said to benefit disproportion- ately those well able to buy books for themselves. While a majority of the public are library members, some estimates are that half of those are accounted for by the 20 per cent of the population labelled middle class (Rusbridger, 1987). Libraries are also accused not only of serving the better off, but also of being elitist, promoting what might be loosely described as middle-class mores which undervalue the cultures of, say, working-class or regional sectors (Dawes, 1978).[2] This prejudice is evident not only in the routine selection of literature which is almost by default 'middle class', but also in occasions of censorship of materials by librarians. In this regard one may point to examples of some libraries removing books such as Enid Blyton's Noddy stories because these have been judged to be racist and sexist. Moreover, the argument is made that behind the rhetoric of public service lies the unpalatable fact that librarians look after themselves rather well, spending only 16 per cent of their budget on books and three times as much on salaries (Adam Smith Institute, 1986: 2). How much better, goes the reasoning, if such a self-serving and elitist profession were made answerable, not to a 'public' which exists only as an abstraction, but to *customers* who, in paying for their information, will value it and call to account those employed to serve it up?

There are other complaints made about public libraries. One is that, since most users borrow light fiction and biographies from libraries (according to the Public Lending Right office these account for around 60 per cent of all loans, with borrowing of fiction amounting to twice that of non-fiction), then there is no reason why their *leisure* pursuits should be subsidised from general taxation, especially since the 'paperback revolution' has made the sort of books that are most heavily borrowed cheaply available.

A second criticism observes a contradiction between public libraries functioning as a free service when it comes to providing information to organisations that want it for commercial reasons. For instance, where a company wishes to investigate a legal or financial matter or to investigate chemical literature as a preliminary to technical innovation, this has consequences of economic significance for businesses yet companies incur no cost in using library resources (and these can be extensive, requiring professional assistance to locate information as well as reference to expensive materials). Critics suggest, with some plausibility, that there is an inconsistency here and that charges should be made in such circumstances.

A third area of concern is public libraries' provision of reference works, probably the area of service that is closest to public service and public sphere ideals. The image is one of the library as a grand repository of 'knowledge', access to which is facilitated by the expert librarian, and of the 'urge to know' of the concerned citizen, the zealous schoolboy, the autodidact, the self-improver, or simply the curious layperson. But against this appealing picture we must set the fact that not only are library reference services not used by a representative cross-section of the public (the better off dominate yet again), but also that reference materials account for only 12 to 15 per cent of library stock and for just 5 per cent of annual book purchases (Adam Smith Institute, 1986: 2). Since most users have enough money to pay their way, and since reference services are a small part of the library's stock, perhaps it is reasonable for free marketers to propose a daily admission charge, with 'season ticket facilities' (ibid.: 36) for longer-term users.

These critiques of public libraries find accord with an enthusiasm for the commercial possibilities of information. Over a decade ago the Information Technology Advisory Panel (ITAP, 1983) published what turned out to be an influential report, revealingly entitled *Making a Business of Information*, which gave voice to this commitment. ITAP urged 'both private and public sectors . . . [to] pay much more attention to information as a commercial commodity' (p.8), advising that entrepreneurs be allowed to enter previously excluded terrain and that those already in position should themselves become entrepreneurial. Public libraries were to the front as recipients of this advice.

Further, ITAP was not advising about a totally new situation. In a key respect it was giving voice to trends that had already been formed, notably the spread of on-line information services. These were from the outset markedly commercial in orientation, aimed consciously at lucrative business markets, in the main. On-line data bases expanded rapidly during the 1980s, chiefly outside the public library system. However, the latter were of course part of the 'information revolution' and were understandably keen to incorporate new forms of information delivery into their repertoires. The problem was that on-line information was both expensive to establish and, at the outset at least, a decidedly minority offering amongst library services. As a result, most on-line information services in public libraries were introduced as an 'extra' for which users were required to pay a fee. As the 'IT revolution' accelerated, so too did the take-up of computer-based information services in libraries – and with this the rapid permeation of a principle (payment for information) at odds with the axial tenet of a free public library service (Lumek, 1984).

What has become evident is that, impelled by additional public demands, by real reductions in resources, by technological innovations and an unprecedented critique of the philosophy underpinning public libraries, a changed conception of information and access to information has emerged. Where once information was perceived as a public resource which ought to be shared and free, now and increasingly it is regarded as a commodity which is tradable, something which can be bought and sold for private consumption, with access dependent on payment. The 'fee or free?' debate is being resolved in an incremental manner in favour of those who favour charging.

It would be wrong to suggest that we have experienced a sea change in the operation of public libraries. New practices are emerging and a new ideology is being articulated (Bailey, 1989), but government continues to exclude charging from basic book borrowing, journals and the use of reference materials (Office of Arts and Libraries, 1988). Nevertheless, 'the levying of charges is gradually becoming more widely accepted' (Lewis and Martyn, 1986), with public libraries charging a fee for inter-library loan requests, for non-book materials, reservation services, out-of-area users, photocopying and, of course, computer-based information.

A genuine fear of these developments is that charges will deter the less advantaged and favour the more affluent and business library user. Charging for services will unavoidably result in the prioritisation of corporate users over individual citizens since the former is more are far and away the most lucrative market. This will be the case especially with regard to information which, privately produced, is prohibitively expensive for individuals (e.g. business consultant reports, many environmental impact assessments, even a good deal of reference materials if they are to be topical). The result of such a tendency must be to leave the general public with reduced access to what one might consider 'hard' information, leading to a 'less informed, less questioning public' (Usherwood, 1989: 18–19) while corporate users enjoy a premium, and economically restrictive, information service (Haywood, 1989).

However, any negative evaluation of such a trend may be met with the observation that, thanks to the declining real costs of information, individuals are in fact in a favoured position to meet the costs of their information needs directly. Indeed, by 1990 the most popular method of obtaining a book was to buy rather than to borrow it from a library. In Britain there are almost as many bookshops as there are public libraries, more titles are published annually now than ever before (in 1986 52,500 new titles appeared, in 1991 there were 68,000), and paperbacks have made books readily accessible to the vast majority of the population. Supportive of this evidence is the growth, over 30 per cent in real terms, of expenditure on books during the 1980s (*Cultural Trends*, 1990a: 14–15). Seen in this light, it is possible to conceive of public libraries as outdated institutions, which once served a purpose in providing information to the public but which have now been made redundant by the development of alternative means of information supply.

There are problems with this line of reasoning. One is that book buyers are heavily concentrated, with over eight out of ten purchases coming from the 25 per cent of the population which is found chiefly in the higher social classes with most education. A second problem is that book buying and library usage are not mutually exclusive. Quite the contrary: heavy users of libraries are also amongst the most likely to buy books. A third problem concerns the types of book people purchase compared with what is offered in public libraries. Much of what people buy is paperback fiction (over 30 per cent of sales), chiefly light 'novels', horror stories, fantasy and thrillers, while non-fiction sales are mainly puzzle books, sports manuals, and DIY publications such as cookery and repair books (Book Marketing Ltd, 1992: 99). Now it is true that public libraries have been criticised for offering too much pulp fiction for free, but they also offer a great deal more than this, especially in the realm of reference works. Use of these is particularly hard to quantify since they are not subject to borrowing, but we do know that standard reference works – from encyclopaedias to gazetteers, statistical sources to business guides – are, as a rule, far too expensive and too frequently appearing in new editions for purchase by individual users. Without public libraries it is hard to imagine people getting ready access to sources such as *Who's Who*, or a legion of year books on subjects as diverse as educational institutions, charitable organisations and political affairs (Ignatieff, 1991). Without public libraries the informational environment of citizens would be significantly impoverished.

A recent report *(Cultural Trends, 1992a)* observed that public libraries in Britain are in decline with fewer books being borrowed while purchases of books by individuals are being sustained in spite of the recession. *Cultural Trends* concluded that this 'is an inevitable result of problems of accessibility, waiting time and limited loan period, together with static or decreasing book stocks and reductions in opening hours. With regard to choice, there are simply fewer books that people want to borrow on the shelves of public libraries' (1992a: 26). It is this sort of evidence that persuades one that the public library network, seen as a foundational element of the public sphere, is being diminished. Fundamental principles, most importantly free access and a comprehensive service, are under challenge, threatened by a new definition of information as something to be made available only on market terms. As this conception increases its influence, we may expect to see further decline of the public service ethos operating in libraries (users will increasingly be regarded as customers who are to pay their way) and with this its public sphere functions of provision of the full range of informational needs without individual cost.

Museums and art galleries

Robert Hewison (1987) concludes his polemical review of changes in museums and art galleries thus:

> In the nineteenth-century museums were seen as sources of education and improvement, and were therefore free. Now they are treated as financial institutions that must pay their way, and therefore charge entrance fees. The arts are no longer appreciated as a source of inspiration, of ideas, images or values, but are part of the 'leisure business'. We are no longer lovers of art, but customers for a product.
>
> (p.129)

Hewison's account of the substitution by the 'heritage industry' of long-established principles of museum and art gallery organisation echoes several themes already reviewed in considering the decline of the public sphere in broadcasting and library provision. So is it possible to understand changes in museums and art galleries with reference to the concept of the public sphere? While I do not think anyone can argue convincingly that these institutions were ever a fully formed public sphere (so many were exclusionary, elitist and intimidating), one can conceive of them as, in important ways, approximations to the ideal. They have several public sphere features:

- The principle of *free entry* to the 'palaces of enlightenment' (as the Victoria and Albert [V&A] was described at its foundation) has long been axial to the operation of British museums and art galleries. This tenet stems from the idea that these institutions have essential cultural and educational functions to fulfil and that, accordingly, access should be open to everyone, irrespective of income.

  To be sure, there are any number of critics who will contest what is classified as 'culture' and 'education' (our imperial past, the celebration of Empire, the tidiness of a good deal of military history, high-class portraiture . . .), but we ought not to forget in all this the Enlightenment roots of the museum and art gallery movement. These roots are not to be lightly dismissed. They stressed the gathering and display of knowledge so that people might be able to know themselves and their world the better to exercise some leverage over it. David M. Wilson (1989), Director of the British Museum, notes that on its foundation by Act of Parliament in 1753, its collections aspired to contain the 'sum of human knowledge' (p.13). Today it is fashionable to observe the fatuity of such an aspiration, but we should not forget that consonant with it is the principle of free access of everyone to what is stored so that they may benefit from being enlightened. Professor Wilson continues: 'Our collections are completely open to scholar and amateur alike and . . . only the most frivolous enquirer will be politely sent away' (p.69).

- *Funding* for museums and galleries, if originally from wealthy benefactors, now comes overwhelmingly from the public purse. Hence, for the year 1988–1989, at least 80 per cent of resource for Britain's national museums and galleries came

117

from the Exchequer (*Cultural Trends*, 1990 b: 11). In spite of this – or rather because of this – the collections are independent of partisan economic and political interests.

• An ethos of *public service* pervades museums and art galleries, with curators and other staff upholding a professional commitment to provide and protect the collections in the interests of the public. Whatever room for interpretation there may be here, high amongst the professional ideals of curators is a non-pecuniary interest in developing collections in service of dispassionate scholarship and their preservation for public edification.

Though in principle museums can act as arenas for critical debate, in practice they have not done much to stimulate it (Walsh, 1992). Frequently they have reflected the class prejudices of their originators and patrons, offering up images, for instance, of Britain's past which may easily be viewed as partial. In addition, patrician origins have often married with Arnoldian sympathies to present, in many galleries, representations of art and an ambience of display which are of an exclusionary 'high culture'. For these reasons visitors are by no means a cross-section of the public: fully 60 per cent of those going to the British Museum have a university degree or equivalent (Wilson, 1987: 87).

But we cannot conclude that *all* there is to museums and galleries are class prejudices. Their cultural contribution entails this, but it goes further. They are highly significant, probably essential, ways of displaying a nation's past and present. Further, Arnold's concern for the 'best that is known and thought' is not wholly the conceit of the privileged. We may disagree about what qualifies as the 'best', but pursuit of the ideal ensures inclusion of works of art of quality worthy of universal esteem. Moreover, the great museums, even if bearing an impress of collectors and donors from what might be seen as a restricted social milieu, also contain exhibits which do open the mind to new experiences and spark wonder amongst visitors. In sum the great museums and galleries are profoundly educative institutions, far removed from 'ideological propaganda', testimony to which are the many recollections of childhood visits – and even personal transformations – from adults. The story of an impoverished young man, later the famous author H. G. Wells, visiting the museum in and around Kensington is well known. When one reflects that 26 million people visited the nation's museums and galleries between 1991 and 92 (*Cultural Trends*, 1993), then it is hard not to believe that countless others find there ways of expanding their horizons, investigating issues, building up their knowledge.

If there are features which suggest some of the qualities of Habermas' public sphere, then it has to be said that they have been put under threat in our museums and galleries in recent years. And how odd it has been that the attack has come from an allegiance of opposites, an alliance of radicals and 'enterprise' enthusiasts, who together charge that these institutions are aristocratic and out of touch. Something of the flavour of changes can be discovered in the prevalent language adopted nowadays: visitors are now referred to as 'customers', 'corporate business plans' are routinely created, and measurable 'performance indicators' are at the forefront of attention. Adding to this is government hostility to the idea of state subsidy which, impelled by the strictures of recession, has meant that museums and galleries have been pressed to manage ever-diminishing budgets.

A common response has been two-fold, introduction of entry charges and seeking for the sponsor. The first has direct consequences for access (and indirect effects on exhibitions), while the second unavoidably limits the autonomy of museum and gallery curators. Entry charges were introduced in the mid- to late 1980s in several national museums and galleries in Britain such as the V&A, the Imperial War Museum and the Natural History Museum. Across the board a fall in attendances was recorded, showing declines of as much as 50 per cent on the 1980s when admission was free (*Cultural Trends*, 1993: 36). Conversely, the one major museum to retain a free entry principle, the British Museum, experienced an unfaltering *increase* in visitors (1986, 3.9 million; 1992, 6.73 million), accounting in 1992 for fully 30 per cent of all admissions.

Sponsors have been the other favoured source of funding. Unfortunately, museums and galleries are less appealing to today's sponsor than the live arts (which in turn are dwarfed by sponsorship monies going to sports) – a good deal more prestige is to be gained from support for Glyndebourne than for a fusty exhibition in the Ashmolean Museum.[3] Still more serious is the fact that sponsors do not get involved for altruistic reasons. They decide to support particular exhibitions and/or institutions for *business* reasons. Sponsorship is a variant of advertising, 'a business tool with a sponsor expecting to get something in return for support' (Turner, 1987: 11). Now it is true that corporate sponsors have a wide range of reasons which impel their business strategies, and these may often mean there is a 'light touch' when it comes to the level of the content displayed in the museum or gallery. Nonetheless, light or heavy, the touch distinctly relies on the desire of the sponsor – something seekers after support must court by planning appropriately attractive exhibitions if they wish the seduction to take place (Shaw, 1990).

Dangers of this situation are obvious on a moment's reflection, though too often they can be ignored by the cash-hungry institution. As an art critic, angry at the spectacular rise of sponsorship during the 1980s that turned 'London's public galleries . . . into shop windows and sumptuous advertising malls for arms manufacturers and credit salesmen' (Januszczak, 1986) has observed:

> Sponsors see the art gallery as a relatively cheap, high profile advertising hoarding and they go there to launder their reputations. They naturally support the kind of art which they calculate will reflect well on them; as their influence grows so does the power of their censorship.
>
> (Januszczak, 1985)

The commitment to commercial practices easily leads museums and galleries to compete for customers with out-and-out market ventures such as Madame Tussaud's. This requires a constant search for the exotic, unusual and dramatically attention-grabbing exhibit that will lure the public and it highlights a growing tendency towards the mounting of 'entertainments' in places dedicated to housing art treasures and historical relics. There is, of course, a grey area dividing making exhibitions accessible and their trivialising artistic and cultural works. Many commentators, however, believe that the boundaries have been crossed, and here they point to the paradox of a boom in commercial museums alongside ongoing crises in state-supported institutions.

The paradox is resolved when these ventures are seen as expressions of the *leisure industry*, 'museums' which offer easily-digested and unchallenging *nostalgia* in a Disney style of elaborate sound effects, eye-catching scenery, quick changes of attractions, video games, animatronic dinosaurs, re-created smells and symbols, and above all 'participation' for the paying customers who are urged to 'enjoy' and have 'fun'. To Robert Hewison (1987) these – everything from the burgeoning growth of commercialised stately homes to theme parks such as Nottingham's 'Tales of Robin Hood' – together represent the ascendancy of the 'heritage industry', which threatens to dominate the arena of museums and galleries, presenting audiences with a cosy and mythological 'England as it once was'.

Government information services

The overwhelming mass of what we know about ourselves as a society comes from the government information services. Most of this reaches us through secondary sources like the press and television, but this in no way negates the point that such information originates from government agencies.

A moment's reflection leads one to realise why this should be so – government is the only institution capable of systematically and routinely gathering and processing information on everything from patterns of divorce to infant morbidity, from occupational shifts to criminological trends, because this daunting task requires huge sums of money and, as important, the legitimacy of constitutional government. Consider, for example, the detailed and intimate information which becomes available from the census every ten years and one appreciates the point. Reflect further that government is the only institution capable of gathering systematic information on such sensitive issues as immigration patterns, or the distribution of income and wealth, and its importance as an informational resource becomes especially clear.

Recognising government as the major agency providing us with information by which we may know ourselves – how we are changing, how health patterns are distributed, how families are structured, how households are equipped – it follows that there is a special need for this information to be reliable. If government policies are going to be effective, still more if citizens are going to be able to evaluate and meaningfully participate in the life of their society, then they must have trust in the information which is fundamental to these processes. Imagine, for a moment, if one could not rely on the accuracy of demographic statistics that tell us of life expectancies, birth rates and regional variations within them; if we could not believe data made available on educational trends such as literacy attainments, different pass rates at GCSE between schools and areas, and classroom sizes; if we could not trust in the integrity of statistics on unemployment rates. . . .

Government information services fit readily into the notion of a public sphere, in that reasoned discussion is unimaginable without a secure knowledge base.[4] It is, indeed, hard to conceive of meaningful politics in which reliable statistical information is absent. For this and other reasons, throughout the nineteenth and twentieth centuries there developed the view that accurate and systematically gathered information should be produced by government as a preliminary to political deliberation of whatever complexion. As former Conservative Cabinet Minister Sir Ian Gilmour has said, the ethos and practice has long been that the 'integrity of statistics should be above politics' (cited in Lawson, 1989).

An essential component has been an ethic of public service amongst the government statisticians who gather and make available this information, one which stresses that the information must be scrupulously and disinterestedly collected and analysed. Statisticians must be both politically neutral and profoundly committed to the professional values of precision, scrupulous methodological practice, objectivity, and a steadfast refusal to distort or suppress evidence (Phillips, 1991). Crucially, these 'custodians of facts' (Phillips, 1988) must rate their wards above political partisanship and pressure as well as above the pursuit of profit. They must also endorse the principle of promptly and unconditionally releasing the information for which they are responsible into the public domain. Sir Claus Moser (1980), one-time head of the Government Statistical Services, articulated these beliefs in an address to the Royal Statistical Society.

> The government statistician commands a vast range of national information and it is his duty to deploy this to the benefit of the entire community . . . he must make readily available, with necessary guidance to sources, such information compiled for and by Government as is not inhibited by secrecy constraints . . . these are not peripheral duties. They deserve high priority. The different user communities not only have a 'right' to information collected and provided for public funds; it is in any case an essential part of a democratic society and of open government that available information should be widely circulated and, one hopes, used.
>
> (p.4)

Finally, because it has been regarded as an essential public service, the dissemination of information has traditionally received a substantial subsidy to make publications affordable to the widest possible cross-section of the public. Particularly significant in this regard is HMSO (Her Majesty's Stationery Office) which until 1980 was an 'allied service' directly funded by Parliament with a brief to make widely available government information. Founded in 1786, HMSO is best known for its publication of Parliamentary debates, reports and legislation and, until recently, if HMSO thought 'a document was "in the public interest" it was sufficient justification for its publication' (Butcher, 1983: 17). What constitutes 'the public interest' is of course contestable, but what is important to note here is that the information, once its publication was agreed, was assumed to be worthy of support so that anyone wanting to receive it could do so without serious economic inhibition.

The suggestion here is not that government statistical services of themselves constitute a public sphere. Rather it is that they are a foundational element of any meaningful public sphere, and that principles such as statistical rectitude, public service and ready public access to government information underpin that supportive role. However, two trends in particular have begun to undermine the traditional role of government information services and, by extension, to denude the public sphere itself. I refer here to the tendency towards treating information as a commodity, and to an increased propensity for government – and politicians more generally – to intervene in ways that threaten the integrity of statistical data. Taken together, these developments amount to a 'politicisation of knowledge' (Phillips, 1989) long considered above the political fray, something which inevitably casts doubt on the reliability and rectitude of information once trusted by all shades of political (and other) opinion.

The first shift may be traced to the aftermath of Sir Derek Rayner's (1981) report to the Prime Minister on government statistical services in 1980. Rayner advocated cutting the costs of government information and shifting the onus away from public service towards charging commercial rates for information to those who required it. Characteristic recommendations from Rayner were that 'subsidy of statistical publications should be quickly curtailed', that information for businesses should 'be charged for commercially', and that, while 'more flexible means of enabling the public to have access to figures held in government should be exploited . . . the costs of providing such facilities should be covered by appropriate charges to the individuals or bodies concerned' (*Government Statistical Services*, 1981, Annex 2). Consonant with this was the decision taken in 1980 to make HMSO a 'trading fund' rather than a service of Parliament, thereby encouraging a more market-oriented mission.

Consequences of this treatment of information by market discipline were cuts in government funding and large increases in the cost of materials going to the public. As Bernard Benjamin (1988) succinctly said, 'the general accusation is . . . that the Government wants to publish as little as possible as expensively as possible' (p.2). For example, Alastair Allan (1990) calculates that there was an explosive increase in HMSO prices in the early Eighties, with for instance, at the top end, *Economic Trends* increasing 457 per cent over a three-year period and the *General Household Survey* rising 72 per cent at the bottom end (p.9). This can only mean greater difficulty of access to information for many, and denial of information to other individuals and groups.

Commitment to a market-guided information policy has also led government to encourage either the privatisation of information services or, if that is not feasible, introduction of internal market mechanisms. Here high tech information companies, providing with computerisation greater speed of delivery, more versatile access as well as simpler dissemination, inject 'added value' to primary data and thus make it more vendible. It is increasingly the case that government is contracting with private companies to make information available in electronic form. This can give a more rapid and virtuoso service (to those who can afford to subscribe), but there are very real fears that it will result in making hard copies of information difficult to obtain.

In the United States this process is the most advanced – and from there come some salutary warnings. For instance, between 1983 and 1993 almost all federal agencies had moved to collect and store data electronically. However, today 'scholars and librarians complain that they do not have access to most of those data bases, or that the data can be reached only through expensive, third-party, commercial vendors that repackage and sell the information'. Furthermore, adds the report, (Wilson, 1993) 'material . . . government once distributed gratis on paper is now available only in electronic form, and only for a very high fee', and now researchers complain 'that even when the government itself makes electronic information available, it is prohibitively expensive' (ibid.).[5]

While this marketisation of government information services has given rise to concern, it is the second trend – the propensity for government to intervene in ways that threaten the integrity of the data – which has caused most upset in recent years. This development may be conceived as an assault on the public sphere by motivated sections which manipulate and even manufacture distorted information to further their own ends. As such, statistics are now to seen *not* as disinterested information, but as a tool of government policy. A more profound blow to the public sphere is hard to envisage.

In a particularly well-researched television documentary, *Cooking the Books* (Lawson, 1989), it was alleged that the Thatcher governments throughout the 1980s systematically intervened in government information services in ways which led to their corruption. It discerned three stages of the production of statistics, during each of which there was political manipulation. First, as regards the *commissioning* of statistics, one of the earliest of Mrs Thatcher's actions on taking office was to abolish in 1979 the Royal Commission on the Distribution of Income and Wealth. The upshot is that reliable data on income and wealth in Britain is extraordinarily hard to come by (Townsend, 1981) and informed debate about viable taxation policies made extremely difficult. Connectedly, and supporting the accusation that 'poverty is a word ministers never want to see in any official report' (Phillips, 1990) – to which end information on such a condition was published after 1979 with less frequency than before – in 1988 the means of measuring it that had been composed in the 1960s were abandoned. The Low Income Families series, which showed the number of people living on state benefits, was commonly used as an index of poverty, but in the late Eighties this was replaced by a new measure: Households Below Average Income. Not only did this thwart attempts to track the direction of poverty trends between generations (of obvious benefit to a government in power during the 1980s which was anxious to belittle the 1960s as an era of waste, indiscipline and social failure), but it also set the base date for the new measure at 1981, a year in which there had been a sharp leap in unemployment and poverty, and thus one which flattered later figures.

The second stage, that of the *compilation* of statistics, suggests further political intervention. Most (in)famous have been the twenty or so changes in definition of 'unemployment' government has insisted upon since 1980, almost all of which reduce the numbers that would be achieved on earlier definitions (Phillips *et al*. 1989). Officially there were about 2.8 million unemployed in Britain in early 1994, but had the criteria not been changed, or if the Major government used measures adopted in other countries, the figure would be nearer 4 million.

The third stage is that at which statistics are *published*. Here Hilary Lawson and her colleagues demonstrate a heightened awareness of the importance of 'timing' and 'presentation' either to promote figures favourable to government or to disguise 'bad news.' For instance, royal weddings dominate the news and in this they are especially useful times at which to reveal data that is unhelpful (for example, a report on leukaemia incidence around the Sellafield nuclear power station was released by the Department of Health on the night of such a wedding, too late to get much coverage in the next day's first editions even if space were made available). More dramatically, there have been occasions when government has attempted to 'bury' its own research, the most notorious case being the Black Report into inequalities in health. Initially only 400 copies were printed and circulated in an attempt to keep from wider audiences the damaging data which correlated social class with morbidity and mortality trends.

Journalist Melanie Phillips, perhaps the most assiduous chronicler of these interventions, believes that 'sensitive statistical information is now manipulated and abused almost as a matter of routine' (Phillips, 1990). There is anecdotal evidence of public scepticism about the reliability of some official statistics, certainly about unemployment figures, but also about crime rates and poverty levels, and a serious consequence of this suspicion is that government information across the board comes to be distrusted. At a deeper level, with this distrust we also get an impoverishment of a central element of the public sphere.

INFORMATION MANAGEMENT

This leads us on to a wider terrain of information management. The public sphere has not only been denuded from within by an assault on its public service functions; it has suffered too from envelopment by a more general development of information 'packaging'. We need to enter here into consideration of the emergence of the 'spin doctor', the 'media consultant', and associated practices in contemporary political affairs. Connected to this is the explosive growth in the means of 'persuading' people, much in evidence in politics, but also extending deep into the arena of consumption. In addition, there has been a massive expansion of 'entertainment', something which results in a surfeit of what Herbert Schiller dismissingly terms 'garbage information'. All told, the thesis is that enormous amounts of greatly increased information in the modern age are of dubious value. There undoubtedly is more information about today, but its informational quality is suspect in the extreme. Let us look a little more closely at this argument.

One of the most striking features of the twentieth century, and especially of the post-war world, has been the spread of the means, and of the consciousness of purpose, of persuading people. As Howard Tumber (1996) observes, 'information management . . . is fundamental to the administrative coherence of modern government. The reliance on communications and information has become paramount for governments in their attempts to manipulate public opinion and to maintain social control' (p.37).

It grew its strongest roots in the opening decades of the century when, as recognised by a spate of thinkers – prominent amongst whom were political scientists like Harold Lasswell (1977) and Walter Lippmann (1922) and, most importantly, the founder of modern public relations Edward Bernays (1955) – the growth of democracy, in combination with decisive shifts towards a consumption-centred society, placed a premium on the 'engineering of consent' (Bernays, 1980).

There is an extensive literature on the growth of 'propaganda', later softened into 'public opinion' and later still into 'persuasion', which need not be reviewed here (Robins and Webster, 1989: 53–68). Suffice to say that it became evident early on in this century that mechanisms of control were necessary to co-ordinate diverse and enfranchised populations. In Lippmann's view this meant a 'need for imposing some form of expertness between the private citizen and the vast environment in which he is entangled' (Lippmann, 1922: 378). This expertise would be the province of the modern-day propagandist, the information specialist in whose hands 'persuasion . . . become[s] a self-conscious art and a regular organ of popular government' (ibid.: 248). Note here that in the eyes of Lasswell, Lippmann and Bernays, information management is a necessary and a positive force: 'Propaganda is surely here to stay; the modern world is peculiarly dependent upon it for the co-ordination of atomized components in times of crisis and for the conduct of large scale "normal" operations' (Lasswell, 1977: 234).

Propaganda here is presented as systematic and self-conscious information management and as a requisite of liberal democracy. It involves both dissemination of particular messages and also the restriction of information, an activity including censorship. What is especially noteworthy about this is that Jürgen Habermas regards the growth of 'information management' as signalling the decline of the public sphere. Habermas is undeniably correct in so far as the promotion of propaganda, persuasion and public opinion management does evidence a shift away from the idea of an informed and reasoning public towards an acceptance of the massage and manipulation of public opinion by the technicians of public relations. Propaganda and persuasion are nowadays usually regarded as inimical to rational debate and are seen as forces which obstruct public reasoning. And yet earlier commentators were quite open and candid about their conviction that society 'cannot act intelligently' without its 'specialists on truth', 'specialists on clarity' and 'specialists on interest' (Lasswell, 1941: 63). As Edward Bernays (1980) proclaimed, 'public relations is vitally important . . . because the adjustment of individuals, groups, and institutions to life is necessary for the well-being of all' (p.3).

What is especially ironic about the present is that information management has become vastly more extensive, much more intensive and much more sophisticatedly applied in the 1990s, while simultaneously there has emerged a reluctance to admit its existence. Nowadays a plethora of PR specialists, of advisers who guide politicians and business leaders through their relations with the media, and of degree courses in advertising and allied programmes, all profess instead to be concerned only with 'improving communications', 'making sure that their clients get their message across', and 'teaching skills in activities essential to any advanced economy'. The underlying premiss of all such practices is routinely ignored or at least understated: that they are dedicated to producing information to persuade audiences of a course of action (or inaction in some cases) which promotes the interests they are paid to serve – i.e. to control people's information environments the better to exercise some control over their actions.

While information management took on its major features in the period between the wars, in recent decades its growth and spread have been accelerating dramatically. Consider, for example, the enormous expansion and extension of the advertising industry since 1945 (Sinclair, 1987: 99–123). Not only has advertising grown massively in economic worth, but it has also extended its reach to include a host of new activities, from corporate imagery, sponsorship and public relations, to direct mail promotion. Consonant with this has been a marked increase in 'junk mail' and free local 'newspapers' which frequently blur the divide between advertising and reportage. Alongside such growth has come about a new professionalism amongst practitioners and a notable increase in the precision of their 'campaigns' (from careful market research and computerised analyses, to 'targeted' audiences).

Further evidence of the growing trend towards managing opinion, and something which reaches deep into the political realm, is the dramatic rise of lobbying concerns that penetrate Whitehall to extend the influence of their paymasters. I do not refer here to the press lobby which gets its name from the place where journalists stand to catch MPs leaving the Commons chamber, but rather to those groups whose aim is to influence the political process itself. A key element of this strategy is the hiring of Parliamentarians by interested parties: today some 30 per cent of MPs are paid 'consultants' (Raphael, 1990). Within the category of Parliamentary consultants one sector shows a particularly vivid presence – MPs employed by political lobbying firms and public relations companies. To many this 'growing number of MPs who are paid by public relations companies . . . for what is often a large and undisclosed list of commercial clients' (Raphael, 1989) questions the integrity of the Parliamentary process itself.

I would draw attention to the contribution of business interests to the information environment. Two features are of particular note. The first parallels the recognition by political scientists of the need to manage the democratic process by careful information handling. In the burgeoning corporate sector, during the same inter-war period, there came about recognition that public opinion could and would increasingly impinge upon business affairs. In the United States especially, 'as firms grew larger, they came to realize the importance of controlling the news which they could not avoid generating' (Tedlow, 1979: 15). The upshot was the establishment of publicity departments briefed to ensure that corporate perspectives on labour relations, economic affairs and even international politics were heard. As Edward Bernays observed, 'in addition to selling its [business] products . . . it needed also and above all to sell itself to the public, to explain its contributions to the entire economic system' (Bernays, 1980: 101).

From acknowledgement that any business organisation 'depends ultimately on public approval and is therefore faced with the problem of engineering the public's *consent* to a program or goal' (ibid.: 159) follows an entire panoply of corporate communications. In the modern business corporation the management of public opinion is an integral element of the overall marketing strategy. To this end the likes of Roger B. Smith, Chairman and CEO of General Motors, are clear about the function of their public relations staff: their instructions are nothing less than 'to see that public perceptions reflect corporate policies' (Smith, 1989: 19). These are the premisses which underlie corporate involvement in myriad informational activities: sponsorship, logo design, corporate image projection, advertorials, public relations, courting of political interests, even involvement with educational programmes. The foundational concerns of the corporate sector are also manifest in joint enterprises, in Britain most prominently in the Confederation of British Industry (CBI), founded in 1965 and now routinely regarded as the authoritative voice of the business community, with acknowledged representation at any public forum to do with the state of 'industry'.

An associated phenomenon is the increasing practice of training leading corporate personnel how best to work with and appear on the media. Speaker training, advice on appropriate dress codes for television appearances, and practice interviews using video facilities, are becoming routine in larger businesses.

Furthermore, Michael Useem (1984) documents how corporate structures have resulted in a greater premium than ever being put on what might be called the informational capabilities of corporations and their leading executives. In Useem's estimation this has led to two especially significant developments. The first is the 'political mobilization of business' (p.150) during and unceasingly since the 1970s. Interlocks between corporations have created a basis which allows the corporate sector to participate effectively in politics on a broadly consensual basis, to respond, for instance, to what may be regarded as excessively high tax levels, to too much power vested in labour movements, or to legislation which hinders enterprise and initiative. In the round the 'political mobilization of business', which precipitated a 'public relations offensive' (Tumber, 1993a: 349) that has continued unabated in Britain since the 1970s, is testament to the need for modern businesses to manage not just their internal affairs, but also the politically fraught external environment which impinges on enterprise. The unrelenting growth of the business lobby – with its opinion leaders, significant contacts, business round tables and constant stream of press releases and briefing documents – and increased support for pro-business political parties, free enterprise think tanks, and vigorous backing to bodies such as the CBI is evidence of a heightened awareness and commitment on the part of the corporate sector.

The second feature concerns the characteristics of today's corporate leaders. Increasingly, Useem argues, they are chosen with an acute eye to their communicative skills. What he terms the rise of the 'political manager' (Useem, 1985) puts great onus on the capacity of business leaders to chart their way through complex political, economic and social environments and to think strategically about the corporate circumstances. An essential requisite of such talents is communicative ability: the ability to persuade parties of the rectitude of company policy and practices. With these traits, aggrandised, interconnected, consciousness of generalised interests, and led by able communicators, corporate interests inevitably exercise a powerful influence on the contemporary information environment.

This second feature of business involvement in the information domain returns us to their more mainstream activities. Again, it is during the key inter-war period that we can discern developments which have profoundly affected today's circumstances. In brief, corporate growth led to the supplement of concern with production (what went on inside the factory) with an increasing emphasis on how best to manage consumption. As one contributor to *Advertising and Selling* observed, 'in the past dozen years our factories have grown ten times as fast as our population.... Coming prosperity . . . rests on a vastly increasing base of mass buying' (Goode, 1926).

In response, corporate capitalism reacted to minimise the uncertainties of the free market by attempting to regularise relations with customers. The steady movement of mass-produced consumer goods such as clothing, cigarettes, household furnishings, processed foods, soaps and motor cars, meant that the public had to be informed and persuaded of their availability and desirability (Pope, 1983). The imperative need to create consumers led, inexorably, to the development of advertising as an especially significant element of marketing (Ewen 1976). Seeing advertising in this way, as 'an organised system of commercial information and persuasion' (Williams, 1980: 179), helps us to understand its role in 'training people to act as consumers . . . and thus . . . hastening their adjustment to potential abundance' (Potter, 1954: 175; cf. Marchand, 1985).

It would be presumptuous to assert that this investment in advertising yielded a straightforward return. People, of course, interpret the messages they encounter and, anyway, advertising is but one part of a wider marketing strategy that might include credit facilities, trade-in deals, and the design and packaging of goods (cf. Sloan, 1965). However, what an appreciation of the dynamic and origination of advertising does allow is insight into the business contribution to the modern-day symbolic environment.

Advertising has grown so enormously since the 1920s, in both size and scale, that it is impossible to ignore its intrusion into virtually all spheres of commercial activity (Fox, 1985). It is today an industry with global reach, dominated by a clutch of oligopolies such as WPP (which owns one-time separate giants Ogilvy and Mather and J. Walter Thompson), Saatchi and Saatchi and Young and Rubicam, yet one which intrudes deep into consumer culture. From billboard hoardings, logos on sweatshirts, tie-in television serials, mainstream consumer advertisements, corporate puffery and sports sponsorships to many university professorships (The Fiat Professor of Italian, the Asda Chair in Retailing), all are testimony to the fact that we now inhabit a promotional culture (Wernick, 1991) where it is difficult to draw the line between where advertising stops and disinterested information starts.

Finally, we might observe also that the need to manage wide spheres of corporate activity reminds us how the advertising ethos carries over from selling goods to selling the company. It is commonplace nowadays to encounter messages that banks 'listen', that oil interests 'care for the environment', that international chemical corporations are 'the best of British', or that insurance companies 'cater for each and every one of us.' We may not be quite so alert to the persuasion, but similar sorts of images are sought whenever companies lend support to handicapped children, or to local choirs, or to theatrical tours. As a leading practitioner in this sector of advertising has confessed, the sole purpose of such persuasion is that companies will 'be given the benefit of the doubt and the best assumed about it on *any* issue' (Muirhead, 1987: 86). Having moved to this level of advertising, we can readily understand how corporate attempts to manage consumption easily merge with corporate ambitions to manage wider aspects of the contemporary scene, up to and including political matters.

What have been considered above are major dimensions of the corporate presence in the information domain. It seems to me to be quite impossible to measure precisely, but, observing the spread of advertising *in its many forms*, as well as the expansion of public relations and lobbying, we can be confident in saying that businesses' *interested* information contributes enormously to the general symbolic environment. Directly in the advertisements which are projected on our television screen, indirectly in the influence advertising brings to bear on most media in the contemporary world; directly in the head of the CBI being asked for the perspective of 'industry' by the newspaper journalist; indirectly through the 'Enterprise Education' materials supplied free to primary schools by Marks and Spencer; directly when a company's Personnel Director is interviewed on television, indirectly when the PR wing succours favour through 'hospitality' and Parliamentary consultants to influence legislation and government policies. . . . Precisely because this information is motivated – and its interests disguised – it denudes the public sphere whenever it enters into that arena (education, political debate etc.) and, more generally, it is a corrupting force in the wider information domain where its economic power gives it disproportionate advantage over less privileged groups.

This section opened with a historical review of the growth of information management in the political realm. Here, in the archetypical public sphere, there is most concern for the intrusion of 'packaged' information since when we cannot be confident about what is read or heard political debate surely loses much of its validity. Yet it is in the polity that trends towards the routine management of information appear most advanced (Franklin, 1994).

There are several important dimensions of this phenomenon. One concerns the presentation of political images, issues and events. The judicious handling of press announcements and statistical data has already been mentioned, but matters of presentation go much further than this. The transformation of Margaret Thatcher, under the tutelage of PR expert Gordon Reece and Saatchi and Saatchi advertising, is very well known (Cockerell *et al.*, 1984): her hair was restyled, her voice delivery reshaped, and her style of dress changed to project less harsh imagery. But the introduction of American-style political techniques went much further than this, extending from the production of speeches that featured snappy 'sound bites' (for example, 'There is no alternative', 'The Lady's not for turning') created to fit the evening television's headlines, to the careful selection of venues for appropriate 'photo-opportunities', if possible with logos, slogans and sympathy-inducing colour schemes on display (Tyler, 1987). Again there is the meticulous preparation of settings for political speeches; these being delivered to invited audiences of the politically supportive. As such they are rallies to celebrate an agreed political platform, not public meetings aiming to argue and convince.

More generally, and notably with political leaders, the events are stage-managed for the television cameras, hence the carefully constructed backdrops, the eye-catching bunting, and, of course, the 'spontaneous' applause. Further, on the rare occasions when television is live, it is well known that politicians take the greatest care to maximise the propaganda effect. Concern is not with open and honest debate, but with using the 'live' interview to best 'manage' public opinion. A nice instance of this occurred early in 1984, when Mrs Thatcher undertook a live evening *Panorama* interview. Coincidentally President Mitterand of France was to visit Oxford the same day and he made it known that he would like to meet his British counterpart. Michael Cockerell (1989) reports that the Prime Minister was 'too busy that Monday to meet the President. Not publicized was her main engagement for the day: the Prime Minister had arranged for a lengthy dry run of the *Panorama* interview – with Bernard Ingham, her press secretary, playing the role of Sir Robin Day. Together they would seek to compose the most persuasive responses to the predicted line of questioning' (p.286). With this one begins to understand how leading politicians appear so able to answer 'unprepared' questions, how they manage to have a battery of illustrative statistics ready in their heads, how much priority is given to presentation of ideas to the wider public, how even 'live' television appearances are part of information management.

Of course politicians have long tried to present themselves and their views in the best possible light. However, it is agreed amongst students of politics and media matters that Mrs Thatcher 'represents the culmination of a hundred years of political news management' (Cockerell *et al.*, 1984: 10). During the 1980s information management in the polity became markedly more systematic and sustained (Harris, 1990: 168–11). Moreover, most contrivance was given to television, itself expanding in services and channels, as greater recognition was given to its centrality to political communications (Young, 1989: 428). Significantly in this regard the Labour Party selected, after its 1983 electoral defeat, as its two leaders Roy Hattersley and Neil Kinnock because they were regarded as televisually appealing (in contrast with the old-style public oratory of their unsuccessful predecessor Michael Foot). In this and in other respects the Labour Party too adopted commercial advertising techniques to sell its candidates and its policies. Tony Blair's untroubled succession to the Labour leadership in summer 1994 underscores the point.

Another dimension of information management is intimidation especially, but not only, of television organisations. During the 1980s there was a good deal of this, from a general antipathy towards the BBC because of its state funding and the conviction of some Tories that it was not 'one of us', to direct attacks on coverage of many issues, the bringing to trial of civil servant Clive Ponting for publishing information counter to Ministry of Defence policy late in 1984, to the successful prosecution of Sarah Tisdall early in 1984 for leaking Foreign Office information to the press.[6] There were especially voluble criticisms of the reporting of an IRA road block at Carrickmore late in 1979, of coverage of the American bombing of Libya in 1986, and, perhaps most heated, of reporting of the SAS (Special Air Services) killing of three active – but unarmed – service IRA members in Gibraltar in 1988 (Bolton, 1990).

Part of the same process was the then Conservative Party Chairman Norman Tebbit's initiative in establishing and encouraging a special unit to monitor all news and current affairs programmes on television for anti-Conservative bias. As an *Observer* (9 November 1986) editorial suggested at the time, the 'real purpose' of such pressures 'is to soften up the BBC (and ITN and IRN) . . . [it] has probably already achieved [its] goal: the self censors will be hard at it from now on'.

Intimidation is often supplemented by censorship, and over the last two decades there has been primary evidence of this. The banning in 1988 of Sinn Fein from British television, the clumsy, ultimately ludicrous, attempt to prevent the publication of former MI5 employee Peter Wright's memoirs, and the revelation that *all* news and current affairs staff appointments were vetted by a secret service staff member located in Broadcasting House are major indexes of such processes (Leigh and Lashmar, 1985).

All three features of information management – information packaging, intimidation and censorship – together with government secrecy which is the reverse side of the same coin are especially evident in conditions of crisis. Here nothing is more compelling than circumstances of war and terrorist activity, of which Britain has experience in Northern Ireland since the early 1970s, in the Falklands in 1982, and, with allies, in Iraq in 1991. Each of these has demonstrated that information has become an integral part of the military campaign, not least that which is for domestic consumption, since public opinion can bear decisively on the outcome of a war effort.

In situations where the 'enemy' has limited access to media outlets and where the military goal is pursuit of victory (rather than truth-seeking), opportunities for distortion and dissembling are plentiful and motivations to deceive are easy to understand. So the media are readily regarded by politicians and the military alike as a means of fighting the enemy, hence as instruments of propaganda. In addition, ever since the American defeat in Vietnam and the emergence of the argument that it was lost due to an uncontrolled press and TV corps (Elegant, 1981; cf. Hallin, 1986), there has developed much more self-consciousness about 'planning for war' on the part of the authorities (Robins and Webster, 1986). During the Falklands War restrictions were placed on journalists' access to the theatre of battle and each was allocated a military 'minder' to ensure proper behaviour; more recently this system has been extended to militarily 'accredit' journalists in time of war (i.e. get reporters to agree to censorship:cf. Harris, 1983; Morrison and Tumber, 1988; Mercer *et al.*, 1987).

The ongoing struggle in Northern Ireland reveals routine manipulation of information (Curtis, 1984; Schlesinger, 1987, ch.8), but it was after the Falklands War that information management became markedly more organised (Ministry of Defence, 1983, 1985). A result was a highly effective PR machine in operation during the Gulf Conflict, media coverage of which was unprecedented in scale yet antiseptic in content. The entire framework was built around the Allies' point of view and their terminology, hence we heard much of 'surgical' air strikes and 'pinpoint accuracy' of bombing, but little if anything of human destruction, a presentation of a 'war almost without death' (Knightley, 1991: 5).

The threat of war and insurgency is not an exceptional condition of liberal democracy, but a routine feature. Because of this, preparedness for such circumstances is characteristic of our age, a key consideration of which is public opinion since this can be crucial in the success or failure of any conflict. This preparedness necessarily results in systematically distorted information, information dissemination not to provide knowledge but to advance the interests of military combatants and politicians. It joins with broader patterns of information management to denude the public sphere, thereby narrowing the range of public discussion and debate.

CONCLUSION

What the foregoing amounts to is endorsement of Howard Tumber's (1993b) conclusion that 'the politics of secrecy and misinformation have been a significant aspect of government political behaviour during the 1980s and 90s and a growing professionalisation of expertise in the management of public information emerged during the years of the Thatcher administration' (p.37). We should add to this the contribution of the business sector to information management, along with a reminder of the undermining of public service institutions and values by a combination of political pressure and commercial emphasis.

Attempts to cast all of this in terms of Habermas' public sphere concept encounter at least two objections. The first concerns the point of comparison from which one contends that there has been a decline. If our starting point is the 1880s, then we must surely arrive at different assessments than if we were we to begin with 1950. Moreover, casting a backward glance over virtually anything but a generation or two, initially at least it does seem odd, even bizarre, to suggest that a public sphere in say the late nineteenth century could be somehow superior to the situation pertaining today since then the majority were disenfranchised and huge numbers even lacked the literacy to be able to read reports in *The Times* and *Morning Post*. Can anyone seriously sustain the argument that people in Britain are more impoverished informationally than their forebears in the previous century? Against this suggestion it is surely unarguable that the public sphere is much more accessible today than ever it was before – think, for example, of the ease with which one may participate in debates on radio phone-ins, or of the facilitative role for organising meetings on the telephone, or of the ease with which one may nowadays amass expert informational assistance.

Such trends have to be openly admitted. Yet we cannot ignore the profound changes that have taken place in the information domain – the commodification of knowledge, the assault on public service institutions, the emphasis on persuasion, the escalation of advertising-oriented media – which means that the potential for and practice of information management and manipulation are immensely enlarged. Perhaps this is the paradoxical situation that we must acknowledge: the opportunities for mendacity and routine interference with information have been seized and are much greater nowadays – in this respect the public sphere is undeniably diminishing. At the same time, there are countervailing tendencies that give people the means and desires to extend and participate in a more open public sphere than has hitherto been offered.

The second objection is that any attempt to assess changes in the public sphere must come to terms with the fact that there has been a growth in the range and complexity of information due both to new media and to the impulses from increased education and the demands of voters. The latter contributes to people's abilities to find out information for themselves, even to research and produce it, as well as impelling the closer investigation of what politicians and business organisations may wish to keep secret. A full assessment of information and the public sphere would need to take the measure of these processes, which may counter a decline in other ways.

These are significant objections, but I do not think they are sufficient to make us abandon either the notion of the public sphere or the argument that it is in decline. This is not just for the reasons sketched above, but also because the *ideal* of the public sphere allows us to estimate the shortcomings of dull reality. And what is striking today is how yawning is the gap between what a public sphere could be in terms of its informational content and what is actually offered.

Moreover, what makes the trends detailed in this chapter especially disturbing is the concept's accordance with another ideal type: *democracy*. Indeed, as was observed at the start of this chapter, it is surely the case that, without a vibrant public sphere, democratic processes are decisively flawed. It is truistic to say, but worth saying anyway against Habermas' bleak pessimism which judges democracy something of a sham because the masses are such passive victims of propaganda, that people are creative and that we have seen imaginative uses of new technologies such as PCs, facsimile machines and video cameras to extend information exchanges. To this degree we can agree that such developments have helped realise a democratic potential in information technologies by advancing those areas of life situated between government and the family which we call civil society and within which one may locate the public sphere. Bulletin boards, rapid and cheap communications, and the felicitous use of camcorders can extend and ease the exchange of information and do much to encourage discussion and debate.

Nevertheless, setting this against the commanding heights of the 'information age' – the global networks dominated by government agencies, transnational corporations and international media conglomerates – who can but see them as strugglers against an unrelenting tide bringing more information management, more commodification of information and the increasing ebb of the public sphere?

7

INFORMATION AND RESTRUCTURING: BEYOND FORDISM?

It is widely acknowledged that established relationships are undergoing major change and that the pace of change is quicker than at any time in history. Not very long ago many working-class youths in South Wales and the North-East could confidently (if unenthusiastically) expect to follow their fathers into the collieries. Those jobs, already reduced in the 1960s and 1970s, totally disappeared during the late 1980s. Now occupations, where they are to be found, are either state-created 'govvies' or in areas such as tourism, retailing and personal care. No one really believes we can return to the old certainties.

Politically we had got used to a world divided into two camps. But 1989 saw an end to that, with the collapse of Communist regimes just about everywhere. In the space of a few months what had become, over seven decades, a fixture of the political scene had gone. What is to follow, no one can be sure, but a return to the recent past is unimaginable.

That people become aware of these and other changes largely through media alerts us to the fact that a key feature of upheaval is information and, of course, the technologies which handle, process and act upon it. The mass media themselves have been radically changed by new ways of gathering and transmitting information – from lightweight video cameras which make it possible to access areas once hard to penetrate, to global satellite links which make it feasible to get pictures on screens thousands of miles away in the space of a few minutes. The whole world could watch as the Berlin Wall came down, when Boris Yeltsin resisted a coup attempt in Moscow, and when the former Yugoslavia was torn apart.

But the import of information in current change is much more than a matter of increasing the messages audiences receive. Many new jobs, for instance, are today what one might call informationally saturated, requiring not manual dexterity and effort, but talking, writing and guiding, something illustrated poignantly by those former coalminers now employed in showing visitors around the reconstructions of collieries in industrial museums such as at Beamish in County Durham. There is also a widespread awareness that information technologies are an integral element of the turmoil itself: the application of computers in factory work means we cannot expect much job expansion there and many of the jobs of the future presume familiarity with computerised equipment. Moreover, computerisation accelerates changes in the here and now and promises continuous change and a consequent need for ongoing adaptation amongst the workforce. Further, the extension of telecommunications around the globe means not only that it is easy to contact friends and relations pretty well anywhere in the world provided they are near a phone, but also that economic and political strategies can, and indeed must, be developed and instigated with a sensitivity towards global factors.

Quite how much information and information technologies are causes or rather correlates of the tremendous changes taking place is a difficult matter to judge, but there are no dissenters from the view that change is deep-seated, that it is taking place on a broad front, that it has been accelerating in recent decades, and that information is an integral part of the process.

To some we are amidst a transfer from an *industrial* to a *post-industrial* society; to a good many it indicates the transition from a *modern* to a *postmodern* world; to Scott Lash and John Urry (1987) it represents a move from *organised* to *disorganised* capitalism; while to Francis Fukuyama (1992) it reveals nothing less than the 'end of history', the triumph of the *market economy* over a failed *collectivist* experiment. Each of these scholars endeavours to explain much the same phenomena, though with different emphases here and there, and, of course, strikingly different interpretations of their significance.

Here I want to concentrate on thinkers who may be divided, at least for analytical reasons, into two interlinked camps, one suggesting that the way to understand contemporary developments is in terms of a shift from a *Fordist* to a *post-Fordist* era, the other arguing that we are leaving behind a period of *mass production* and entering one in which *flexible specialisation* is predominant.

In my discussion of a purported transition from Fordism to post-Fordism it is my intention to concentrate on ideas emanating from what has become known as *Regulation School* theory. Here the major originators are economists Alain Lipietz (1987), Michel Aglietta (1979) and Robert Boyer (1990), though I shall incorporate several independent analysts, notably David Harvey (1989b) and Scott Lash and John Urry (1987, 1994) who appear to have a good deal of agreement on major facets of change. As I turn to flexible specialisation theorists I shall focus attention on the most influential publication, Michael Piore and Charles Sabel's *The Second Industrial Divide* (1984).

REGULATION SCHOOL THEORY

Regulation School theory emanated from a group of French intellectuals, themselves influenced, especially early on, by Marxist economic thinking, though several key contributors, notably Michel Aglietta, quickly distanced themselves from such traditions while others such as Lipietz have been increasingly responsive to questions raised by ecological movements. Regulation School theory does, however, retain one element closely associated with at least some Marxist traditions, namely the search for a *holistic* explanation of social relations, which attempts to grasp the overall character of particular periods. In doing so it also lays stress on the ways in which a range of features *interconnect* to enable a society to perpetuate itself.

The fundamental question asked by Regulation School is: how does capitalism ensure its perpetuation? How does a system that is premissed on the successful achievement of profit and consistent expansion of capital achieve stability? Or, to put this in terms Regulation Theory thinkers prefer, how is *capitalist accumulation* secured? Of course, it could be argued that any system which is in a constant state of motion, and capitalism is undeniably one such, is inherently unstable and that therefore there is something odd about Regulation School's search for the roots of stability in a dynamic economy (Sayer and Walker, 1992). Regulation School thinkers concede the point that instability is part and parcel of capitalist relations, but they are also taken with the question: how does capitalism manage to continue in spite of these sources of tension? In other words, Regulation School seeks to identify ways in which instabilities are managed and contained such that continuity can be achieved amidst change.

Regulation School thinkers seek to examine the *regime of accumulation* which predominates at any one time. By this they mean to identify the prevailing organisation of production, ways in which income is distributed, how different sectors of the economy are calibrated, and how consumption is arranged. They also try to explain the *mode of regulation*, by which is meant the 'norms, habits, laws, regulation networks and so on that ensure the unity of the process (of accumulation)' (Lipietz, 1986: 19). This latter, concerned with what might be termed the 'rules of the game', takes us on to consideration of ways in which social control is achieved, from legal statutes to educational policies.

Regulation School adherents aim to examine the relationships between a regime of accumulation and its mode of regulation, but in practice most studies from within the school have focused on the mode of accumulation. Their contention is that, over the last fifteen years or so, the ongoing crises with which we are all more or less familiar (recession, unemployment, bankruptcies, labour dislocation etc.) are being resolved by the establishment of a new regime of accumulation which is to replace the one that secured stability for a lengthy period after the Second World War. The suggestion is that the Fordist regime of accumulation which held sway from 1945 until the mid-1970s became unsustainable and that, hesitatingly and with considerable disruption, it is now giving way to a post-Fordist regime which will, perhaps, re-establish and sustain the health of capitalist enterprise.

In what follows I shall concentrate on contrasting the Fordist and post-Fordist regimes of accumulation. This will, inevitably, be at the expense of much attention being given to modes of regulation, but readers ought to be aware of that omission (Hirsch, 1991). Particularly as they read of attempts to construct a post-Fordist regime during the 1980s they might reflect on the control mechanisms that were introduced in Britain during those years, from Mrs Thatcher's determined assault on the labour movement and radical revisions of the structures and syllabuses of schools and higher education, to reorganisation of local government (Gamble, 1988; Kavanagh, 1990).

FORDIST REGIME OF ACCUMULATION, 1945–1973

Regulation School theorists contend that these years may be characterised as the *Fordist–Keynesian* era, during which a number of interconnected features ensured that the system as a whole maintained equilibrium. Briefly, this was an expansionary period in which mass production and consumption were in reasonable balance, in which state involvement in economic affairs helped keep that harmony, and in which government welfare measures assisted in this as well as in upholding social stability.

Because Ford was the pioneer of production techniques which allowed the manufacture of goods at a price that could encourage mass consumption while he was also at the forefront of payment of (relatively) high wages which also stimulated the purchase of goods, his name has been applied to the system as a whole. However, it would be an error to suppose that Ford's methods were established either everywhere or in the same way.[1] Rather the terminology indicates that the Ford corporation was the archetype, especially at its peak in the post-Second War phase when it came to represent many of the key elements of advanced capitalist enterprise. Similarly, since Keynes is the economist whose policies are most closely associated with state intervention in industrial affairs, the term Keynesian should be understood paradigmatically rather than as suggesting that governments acted in a uniform manner across different nations.

The Fordist–Keynesian era had a number of important distinguishing features:

1 *Mass production* of goods was the norm. Here, in areas such as engineering, electrical goods and automobiles, it was characteristic to find standardised products, manufactured using common processes (the assembly line system), being created in large volume in pretty much undifferentiated patterns (fridges, Hoovers, televisions, clothing, etc.). Typically manufacturing plants were large, at the upper end the Ford factory in Detroit having 40,000 employees on the one site, but even in England the Rover car plant in Oxford had some 28,000 workers in the late 1960s, and, since everywhere cost-effective mass production required the economies of scale which came with size, factories of several hundred or even thousands of employees were typical. In the United Kingdom by 1963 fully one-third of the entire labour force in private sector manufacture worked for organisations with at least 10,000 on their pay-roll and over 70 per cent of people in manufacture worked in companies with more than 500 employees (Westergaard and Resler, 1975: 151–152). A corollary was the development of distinctive localities known by what they produced: for example Derby for its railway works and Rolls-Royce factory, Shotton, Corby and Consett for their steelworks, Coventry for its automobiles, and Birmingham for various engineering enterprises.

2 The predominant group in employment was *industrial workers*. These were the male, blue-collar employees employed in manufacture and some extractive industries who evidenced strong regional and class attachments which were echoed in political affiliations and attitudes. Constituting almost 70 per cent of

the British workforce in 1951, male manual workers still accounted for almost 60 per cent of the total twenty years later (Harrison, 1984: 381) and, in the early 1960s, about 60 per cent of all employment was located in sectors covering a range of industrial activities from mining to chemical production, while 43 per cent of jobs were accounted for by manufacturing alone (Gershuny and Miles, 1983: 20).

In industry there was a high degree of unionisation amongst the workforce, which was recognised by most employers and channelled into institutional arrangements for handling labour and management relationships. At the local level this found expression in agreed negotiation procedures while at the highest levels it was reflected in a tendency towards corporatism (Middlemas, 1979) by which employers' representatives, trade union leaders and politicians would meet regularly to agree on issues of mutual concern.

Above all, perhaps, the longest boom in capitalism's history meant continual economic growth and, with it, full employment.[2] With the exception of a few areas, unemployment in Britain virtually disappeared, rates hovering around 2 per cent throughout the 1950s, something which brought stability, assurance and confidence to the majority of the population.

3 Over these years *mass consumption* became the norm, facilitated by (relatively) high and increasing wages, decreasing real costs of consumer goods,[3] full employment, the rapid spread of instalment purchase[4] and credit facilities, and, of course, the stimulation that came with the growth of advertising, fashion, television and cognate forms of display and persuasion.

People gained access to hitherto scarce and even unimagined consumer goods – from toiletries and personal hygiene products, stylish and fashionable clothing, vacuum cleaners, fitted carpets, refrigerators, radios and televisions, to motor cars – in the years following on from 1945. By 1970 nine out of ten homes had a television, seven out of ten a fridge, and over six out of ten a washing machine, while car ownership rose from 2.3 million in 1950 to 11.8 million in 1970, so that over half the nation's households had an automobile (Central Statistical Office, 1983, table 15.4: 278).

Most important, mass consumption relied on the working class gaining access to what was offered since it was they, being the majority, which constituted the biggest market for goods. More than this, *mass consumption became an axis of continuous and stable mass production.* During this epoch steady and sustained mass consumption of goods was a requisite of an expanding production base which in turn ensured full employment. During the Fordist era the health of the economy was increasingly determined by the strength of consumer purchases. It became a virtue to consume.

The crucial point is that some calibration was achieved between mass consumption and mass production. To ensure that this continued a whole edifice of marketing and design techniques was developed – annual model changes in cars, a burgeoning advertising industry, new layouts of shops, trade-in deals, easy terms for purchase – but surely most important was the assurance of full employment and continuous real increases in income. So long as consumer demand was strong (and the state intervened frequently to ensure that it was), the economy could remain vibrant.

4 Throughout this period the *nation state* was the locus of economic activity and within it sectors were typically dominated by a cluster of *national oligopolies*. Characteristically, three or four dominant companies would be identified in any one area, be it electronics, clothing, retailing or engineering. In line with this, in 1963 the leading five businesses in British manufacture accounted for almost 60 per cent of all sales in any trade area (Westergaard and Resler, 1975: 152). More generally, the top one hundred companies achieved one-third of all Britain's manufacturing output in 1960, underlining the hold of large corporations. Indigenous companies had a firm hold on the domestic market: as late as 1968 manufacturing industry was 87 per cent British by output.

With hindsight we can see that British industry was rather comfortably situated. It controlled most of the domestic market, it had few competitors, it was participating in steadily growing and secure markets and, increasingly, it was vertically and horizontally integrated so that it could maximise control and co-ordination over its interests.

5 Underpinning much else was an acknowledged role for *planning* (Addison, 1982), most vividly manifested in the growth of the Welfare State, but also expressed in a broad consensus on the legitimacy of state involvement in the economy. Significantly, for example, the tide of nationalisations that followed the Second World War and took over much energy supply and communications was turned back by the Conservatives only in the steel industry during the 1950s. Other areas such as coal, gas and electricity were accepted across the party divide. The suggestion of Regulation School theorists is that this sort of accord bolstered extensive planning in many areas of life, as well as winning support from most people, who felt that state-supplied education and health especially were of great benefit to themselves, thereby helping maintain stability throughout the Fordist system.

One may gain a better appreciation of the depiction of a Fordist regime of accumulation by taking into account some of the major social and economic trends and events of the 1970s. It was at this time that, amidst a sharp recession and the shock of large-scale oil price rises in 1973,[5] that there came about an awareness that developments were taking place that meant the Fordist regime was no longer sustainable. Post-Fordism, signalled by the trends which undermined Fordist conditions, began to emerge during this period. At the storm centre of these changes were ways of handling, storing and acting on information.

GLOBALISATION

One of the most important factors that led to the downfall of Fordism and which is often thought of as characteristic of the post-Fordist era, is globalisation. This is a long-term development, and still far from accomplished, but it accelerated during and since the 1970s. The term refers *not* merely to an increasing internationalisation of affairs which suggests more interaction between autonomous nation states. Globalisation means something much more than this: it signals the growing interdependence and interpenetration of human relations alongside the increasing integration of the world's socio-economic life. There is a tendency to conceive of globalisation as primarily an economic affair, manifest in the tying together of markets, currencies and corporate organisations. It is this, but it is simultaneously a social, cultural, and political condition evident in, for example, an explosive growth of migration, of tourist activity, hybrid musical forms, and heightened concern for global political strategies to meet threats and challenges to survival.

Capitalism, the social form which has pioneered globalisation, has proven itself extraordinarily successful: it has extended its reach across the globe simultaneously with penetrating deep into intimate realms of life. Capitalist activities are today at once world-wide (and rapidly extending into hitherto isolated areas such as the former Soviet Union and China) and, at the same time, well able to enter into spheres such as childcare, personal hygiene, and provision of everyday foodstuffs. As it has done this, capitalism has brought the entire world into networks of relationships such that, for example, we may get our coffee from one part of the world, our wines from another, they their television from one region and their clothing from another, all of this conducted by interconnections which integrate the globe. Quite simply, the trend is towards the world being the context within which relationships are conducted, no matter how localised and particular an individual life may appear to be experienced.

In addition, and crucial to the operation of globalisation, is the expansion of transnational corporations (TNCs) that have provided the major foundations of this phenomenon. It is important to appreciate the especially rapid growth and spread of transnationals in recent decades. For example, in 1950 only three of the 300 largest TNCs had manufacturing subsidiaries in more than twenty countries, but by 1975 forty-four TNCs from the United States alone had that presence. Again in 1950, the vast majority of American TNCs had subsidiaries in fewer than six countries; by 1975 only nine operated on such a small scale (Dicken, 1992: 50).

The size and scope of TNCs is hard to grasp, but some idea might be gauged by noting that, when the wealth of nations and corporations is scaled, TNCs can account for half of the largest one hundred units. In fact, in financial terms only a couple of dozen countries are bigger than the largest TNC. The likes of General Motors (1992 revenue $133 billion), IBM ($65 billion), Shell ($99 billion) and General Electric ($62 billion) are indeed 'the dominant forces in the world economy' (Dicken, 1992: 49) and they account for as much as 25 per cent of total world production (ibid.: 48). They are themselves highly concentrated, the biggest of the TNCs accounting for the lion's share of activity in any given sector. For instance, Peter Dicken (1992) identifies a 'billion dollar club' of just 600 TNCs which supply more than 20 per cent of total industrial and agricultural production in the world's market economies, yet within these giants 'a mere seventy-four TNCs accounted for fifty per cent of the total sales' (p.49).

Globalisation, operationalised and constructed by transnational corporations, has a number of especially significant features. Prominent amongst these are the following.

Globalisation of the market

This means that the major corporate players now work on the assumption that their markets are world-wide and that these are now open to all economic entities with the resources and will to participate in them. Of course, even nowadays few TNCs operate with a pure global strategy, but this is undoubtedly the direction in which they are moving.

Globalisation means that markets are today bigger than ever before and that increasingly they are restricted to those with the enormous resources necessary to support a global presence (Barnet and Müller, 1975). Paradoxically, however, markets are in key respects *more fiercely competitive* than previously precisely because they are fought over by giant corporations with the resources to have a global reach. At one time a national market might have been dominated by a local oligopoly, but, over the years, these have increasingly been trespassed upon by outsiders (and, of course, energetic indigenous corporations have themselves moved outside their home country to attack other markets). These new challengers, in establishing a global presence, are at once bigger and more vulnerable than hitherto. Look where one will and one sees evidence of this process; for instance the motor industry now operates at a global level, with vehicles being marketed on a world scale, which means that one-time national champions can no longer be secure, a point underlined by the takeover in 1994 of the last major British motor vehicle manufacturer, Rover, by BMW (though Honda already owned 20 per cent of the 'British' outfit). Much the same features are manifest in petrochemicals, pharmaceuticals, computers, telecommunications equipment and consumer electronics. In fact, virtually everywhere nowadays the market is increasingly a global one.

This world market is roughly divisible into three major segments – North America, Europe and the Far East – but of course the major TNCs operate extensively in all three domains. This tripartite division reminds us of something else that globalisation of the market means. I refer here to the emergence in little more than a generation of what are today perhaps the archetypical global corporations, namely Japanese conglomerates which frequently profess to having no national roots (other than in those countries in which they happen to invest). The likes of Hitachi (1992 revenues $61.5 billion), Matsushita ($57 billion), Toyota ($79 billion) and Sony ($31.5 billion) have distinctive global strategies for their product ranges. Over the years, in automobiles, consumer electronics and, most recently, information technologies these have proven to be highly successful. Simultaneously they have contributed to a massive shake-up of established corporate interests in the West. Whether in automobiles, office equipment, televisions, video or computers, the Japanese challenge has rocked what was, at least for a time, a comparatively settled economic order.

Globalisation of production

It follows that, as corporations are increasingly involved in global markets, then they must arrange their affairs on a world scale. Global production strategies are a central feature of such a development, TNCs more often arranging, for example, to locate their headquarters in New York City, design facilities in Virginia and manufacture in the Far East, with sales campaigns co-ordinated from a London office. The inexorable logic of globalisation is for TNCs to plan for such strategies in order to maximise their comparative advantage.

This development catapults informational issues to the fore, since how else can market strategies and world-wide manufacturing facilities be organised other than with sophisticated information services? I shall have more to say about this later, but here may observe that the globalisation of production also encourages the growth of what Dicken (1992) calls 'circulation activities' which 'connect the various parts of the production system together' (p.5). That is, an essential condition of the globalisation of production has been the *globalisation of information services* such as advertising, banking, insurance and consultancy services which provide 'an emerging global infrastructure' (ibid.). For instance, American Express, Citicorp, BankAmerica, Lloyds and Merrill Lynch also straddle the globe, servicing the corporate industrial outfits which they closely parallel in their structures and orientations.

Globalisation of finance

A central aspect of globalisation is the spread of world-wide informational services such as banks. These suggest something of the globalisation of finance, but it also refers also to something much more: nothing less than the development of an increasingly *integrated global financial market*. With sophisticated IT systems now in place, plus the deregulation of stock markets and the abolition of exchange controls, we have nowadays facilities for the continuous and real-time flow of monetary information, for round-the-clock trading in stocks, bonds and currencies. These developments have enormously increased both the volume and velocity of international financial transactions, bringing with them a heightened vulnerability of any national economy to the money markets.

The scale and speed of these informational flows is astonishing. Will Hutton (1994), for instance, observes that foreign exchange turnover now dwarfs the size of national economies and makes trade flows (a traditional method of measuring national economic activity in terms of import and export levels) appear small in comparison. Thus 'the total level of world merchandise trade in 1993 is two-thirds of US GDP; it will take turnover in the foreign exchange markets less than a fortnight to reach the same total – leaving aside the cross-border derivative, bond and equity markets' (p.13). Joyce Kolko (1988) traces an exponential growth in foreign exchange trading, which doubled between 1979 and 1984 and put on another 30 per cent in the following two years (p.193; cf. Coakley, 1992). More recently, *Fortune* magazine (26 July 1993) reports that flows through the US-based Clearing House Interbank Payments System average $850 billion or more per day and sometimes pass $1 trillion (p.26); this had grown 400 per cent over the last decade. Along similar lines, a *Financial Times* journalist, observing the dramatic increase in the foreign exchange market to $1,000 billion per day in 1993, reports also that cross-border trade in equity holdings rose from $800 million in 1986 to $1,300 billion in 1991 (Blitz, 1993). Again, consider the spectacular growth of trade in Eurodollars.[6] This trade grew from but a few hundred million dollars, handling issues of between $2 and $5 million each in the mid-1950s, to one worth $300 billion in 1985, handling issues of between $200 million and $500 million a time (Hamilton, 1986: 21).

Globalisation of communications

Another dimension of globalisation is the spread of communications networks that straddle the globe. Clearly there is a technological dimension to this – satellite systems, telecommunications facilities and the like – to which I shall return, but here I would draw attention to a phenomenon discussed in the two previous chapters: the construction of a *symbolic environment* that reaches right around the globe and is organised, in very large part, by media TNCs.

This has many important social and cultural consequences, but here I emphasise only the bringing into being of an information domain which provides people with common images. For instance, in 1991 a core of movies, originating in the United States, achieved far and away the largest audiences wherever they were shown across the globe. *Dances with Wolves, Terminator 2, Robin Hood Prince of Thieves*, and *Silence of the Lambs* were box office leaders in Germany, Britain, Italy, France, Spain, Australia, the USA – pretty well everywhere that there were cinemas. This provides audiences, widely diverse in their responses and dispositions though they be, with a mutual symbolic sphere – and much the same could be said about today's television shows, news agencies or, indeed, fashion industries.

However much one might want to qualify statements about just what consequences this might have when it comes down to particular people and particular places, this globalisation of communications has a significant part to play in the functioning of the global economic system. To be sure, it cannot be said unequivocally that American television soaps dispose viewers towards the lifestyles portrayed, that the advertisements carried successfully persuade, that the designs displayed in the movies stimulate yearnings amongst audiences, or that the rock music emanating from Los Angeles and London encourages the world's youth to seek after the styles of clothing of and foods eaten by its performers. It is unarguable that these global images often incorporate several elements of different cultures so they are not unidirectional in their orientation. But what is surely impossible to dismiss is the view that it is not possible to imagine large parts of the world's economic forces continuing without the underpinning of this symbolic milieu. It may not be sufficient in itself to persuade, but it is necessary to most commercial endeavour.

Each of these dimensions of globalisation requires and contributes towards an information infrastructure to cope with the changed stresses and strains of world-wide operation. That is, as globalisation grew and as it continues, so ways of handling information and information flows have been put in place. We can identify major elements of this informational infrastructure.

- The world-wide spread and expansion of services such as banking, finance, insurance and advertising are essential components of globalisation. Without these services TNCs would be incapable of operation. Information is, of course, their business, the key ingredient of their work: information about markets, customers, regions, economies, risks, investment patterns, taxation systems and so forth. These services garner information and they also generate and distribute it, having added value by analysis, timeliness of response or collation.
- Globalisation requires the construction and, where necessary, enhancement of computer and communications technologies. In recent years we have seen the rapid installation and innovation of information technologies which are a requisite of co-ordination of global enterprises.
- This information infrastructure has resulted in the growth of information flows at a quite extraordinary rate. For instance, business magazine *Fortune* (13 December 1993: 37) reports that international telephone connections to and from the United States grew 500 per cent between 1981 and 1991 (from 500 million to 2.5 billion). Elsewhere, there has been an astounding expansion of financial traffic along the international information highways. Exchange rate trading, direct foreign investment patterns and the markets in bonds and equities have expanded apace, underlining the import in global markets of the flows of financial information.

THE DEMISE OF FORDISM?

Globalisation has meant that Fordism is increasingly hard to maintain. How could things be otherwise when Fordism's organisational premiss – the *nation state* – is undermined by the international spread of transnational corporations and the constant flow of information around and across the globe? Fordism hinged on the sovereignty of nation states, on governments' capacity to devise and implement policies within given territories, on the relative immunity from foreign competition of indigenous companies and on the practicality of identifying distinctively national corporations. But these conditions are growing more rare in the days of global marketing, frenetic foreign exchange dealings and enterprises located at multiple points around the world.

It would be foolish to deny that the nation is important for a great many aspects of life, but economically at least it is becoming less relevant. There are two particularly significant indications of this. The first is that the rise to prominence of transnational corporations obscures what is owned by any given nations. To what extent, for example, can one consider GEC or Hitachi a specific nation's property? Corporations such as these are usually given a national label, but with very large proportions of their production and investments abroad it is difficult to unambiguously designate them British or Japanese. For instance, as early as the 1970s in Britain over 50 per cent of manufacturing capacity in high technology (computers, electronics etc.) and heavily advertised consumer goods (razors, coffee, cereals etc.) was accounted for by subsidiaries of foreign firms (Pollard, 1983). Are industries located in the UK such as Nissan (Sunderland), IBM (Portsmouth) or Gillette (London), British, Japanese or American? An investigation in *Labour Research* (1983) found that 44 per cent of the output of Britain's top fifty manufacturing companies took place overseas. A disturbing supplementary question follows: to whom then are these TNCs responsive? If they have substantial investment outside the jurisdiction of what one might think of as their 'state of origin', then to whom are they answerable? That begs the question of ownership, a matter of considerable obscurity, but we can be confident, in these days of global stock market dealings, that TNCs will not be owned solely by citizens of any one nation. To the extent that private corporations remain responsive primarily to their shareholders, then this international ownership necessarily denudes conceptions of the 'national interest' and strategies developed by particular nation states.

A second way in which the nation state, and thereby Fordist regimes, is undermined is by pressures generated by operating in a global economic context (cf. Sklair, 1990). If nation states are becoming less relevant to business decisions as investors and TNCs seek the highest possible returns on their capital around the world, then individual countries must encounter overwhelming pressures to participate in, and accord with, the global system. This is nowhere more acutely evident than in the realm of financial flows, with nation states nowadays especially vulnerable as regards currencies and investments should governments attempt to do anything out of line. The integration and interpenetration of global economies has resulted in nations having to shape themselves in accord with international circumstances, the upshot of which is that individual states 'have found it extraordinarily difficult to maintain their integrity in the face of the new international realities of capitalism' (Scott and Storper, 1986: 7).

Most nations now seek, more or less avidly, investment from TNCs, but the necessary precondition of this is subordination to the priorities of corporate interests which are committed to market practices (in so far as these maximise their interests) and at the same time are not restricted to particular territories.

The outcome of unification of the world's financial markets has been that individual governments find their monetary sovereignty challenged whenever investors and traders sense vacillation or weakness, something experienced in the early 1990s in Britain, Ireland and Spain as well as in other countries. This means that political options and the autonomy of governments are taken away, since

> an anonymous global capital market rules and its judgements about govern-
> ments' credit-worthiness and sustainability are the ultimate arbiter – and
> much more important than the opinion of national electorates. It is before
> these that so many governments quail. If they do not obey the . . . policies
> that the market approves, then their debt and currencies will be sold – forcing
> them to face an unwanted policy-tightening.

> (Hutton, 1994)

During the mid-1960s the then Labour Prime Minister Harold Wilson complained of mysterious 'gnomes of Zurich' whose trading in sterling compelled his government to devalue the pound and reduce public expenditure. These experiences are frequently cited as instances of the power of financiers to limit national policies. And so they are, but how much more inhibiting are the pressures of today's immensely more integrated, electronically connected, financial centres.

POST-FORDISM

These trends, combined with the recessions which afflicted advanced capitalism during the 1970s, have stimulated the creation of a new regime of accumulation. The suggestion is that, after a twenty-five-year period of stability, Fordism had run its course. New circumstances required radical changes, not least a thorough restructuring of corporate organisations if they hoped to achieve the sustained expansion once enjoyed and come to terms with the new milieu in which they found themselves.

An important part of this was to be an assault on organised labour, initially the trade unions, but extending to collectivist ideas *tout court*, especially in Britain and the United States. At one level labour needed to be attacked because its traditional practices were an obstacle to any deep-seated change, but at another it was symptomatic of the more generally cumbersome and entrenched character of the Fordist era. Globalisation and continuing economic uncertainty demanded, as we shall see, rapidity and versatility of response, things which Fordism's set and stolid ways cannot deliver.

A requisite of profound change was therefore an industrial relations policy which disempowered the trade union movement. In the United States this was relatively easy, and after President Reagan's defeat of air traffic controllers in the early 1980s there was little resistance to change. In Britain there was a more formidable labour movement, but it too was defeated by a variety of means: from legislation that weakened the effects of pickets and increased the financial liability of unions in law and a willingness to tolerate unprecedentedly high unemployment which grew over 200 per cent between 1979 and 1981 and cut a swathe through manufacturing industry where were found the most organised working-class jobs, to a very determined government which defeated attempts – notably by the miners in a long strike during 1984 and 1985 – to thwart proposals to radically change their industries and occupations.

A close correlate was moves to shed labour, a necessary corporate response to stagnant markets, but of longer duration in two respects. One, which is euphemistically termed 'downsizing', has continued over the last decade and more, with many successful corporations proving themselves able to generate 'jobless growth'. So a common feature of the post-Fordist regime has been a capacity to increase productivity by either or both extra effort from employees and the application of new technologies on such a scale that economic expansion is combined with labour reductions. This is by no means a universal trend, but many examples can be found, for example IBM has shed 25 per cent of its 400,000 labour force since 1983 though its income about doubled; by late 1993 British Telecom had lost 100,000 of its 1989 complement of 250,000 while over the same four-year period revenue grew by over a fifth; and British Petroleum got rid of over a quarter of its 132,000 staff between 1983 and 1993 while income rose over 20 per cent.

The second feature is more often regarded as a distinguishing aspect of post-Fordist organisation. The suggestion is that corporations have increasingly begun to vertically disintegrate, by which is meant that, instead of producing as much as is possible within the single organisation (and hence endeavouring to be vertically integrated), there is a trend towards contracting with outsiders for as many as possible of the company's requirements. This strategy of *outsourcing* fits well with downsizing since it requires relatively few employees in the central organisation and helps when it comes to redundancies (contracts are not renewed instead of staff being sacked). Benetton, the Italian clothes manufacturer, is the usual reference here (Murray, 1985), an outfit which uses 12,000 workers to produce the apparel, but has only 1,500 in direct employment. Benetton's strategy of franchises (it has over 3,000 in 57 countries) is outsourcing, a route which releases the corporation from the responsibility of keeping large numbers of permanent employees on its books.

Vertical disintegration is feasible only when there is an adequate *infrastructure of communications and computer facilities* of sufficient sophistication to allow the co-ordination and control of dispersed activities. How else could Benetton's 140 or so agents, each with a designated geographical region for which they are responsible, co-ordinate affairs? This infrastructure is regarded as an essential component of post-Fordism for several reasons, all of which underline the heightened role of information in the new regime. I have already drawn attention to aspects of it in the discussion of globalisation which presaged post-Fordism, but several features of the information infrastructure may be highlighted.

1 It is essential to allow the orchestration of globalised production and marketing strategies. Several commentators propose that, in recent years, we have witnessed the spread of a *new international division of labour* (cf. Fröbel *et al.*, 1980), one overseen by transnational corporations capable of managing production, distribution and sales world-wide, co-ordinating sites in dozens of

international locations. Just as outsourcing depends upon computerised communications which enable organisations to achieve continuous observation of suppliers and distributors without employing large numbers of staff in-house, so too is a global corporate strategy feasible only on the basis of a sophisticated information network. Furthermore, the restructuring process to which we alluded above, in all its dimensions, but especially in its 'global option' (shift production to Manilla, component supply to Prague, enter markets in Moscow and get some facilities in Cork . . .) 'would have been inconceivable without the development of information technologies, and particularly telecommunications' (Henderson, 1989: 3). This is, of course, why data traffic over phone wires is growing by 30 per cent every year in the USA (*Fortune*, 13 December 1993: 35).

2 It is crucial to the handling of the global financial trade and cognate information services, which are essential components of a globalised economy. Without reliable and robust information networks the extraordinary volume and velocity of share trading, stock market exchanges, inter-bank and bank-to-client communications, plus associated activities, would be untenable, and so, by extension, would be the post-Fordist regime of accumulation.

3 It is central to improvement of products and production processes, offering not just greater effectiveness and efficiency by providing more precise monitoring and thus better control functions, but also frequent opportunities to introduce new technologies that are cost-effective and/or enable improvements in quality (one thinks here of the ongoing automation and mechanisation manifested in robotic applications, computer numerical control and general computerisation of office work).

4 It is an integral element of endeavours to enhance competitiveness in an ever-more intensely rivalrous context. To stay abreast, still less ahead, of the competition it it essential that companies are to the forefront of new technologies. But the pressure to improve one's competitive edge extends to much more than having state-of-the-art computerised technologies on the shop floor. It is just as important that one's networks are developed and used optimally – within the organisation and between organisations, so that efficiency will be increased; to and from one's subsidiaries and suppliers, so that weaknesses may be eradicated and strengths built upon; and to one's markets, so that opportunities can be seized. Increasingly it appears the successful corporation is that which is highly automated on the shop floor and offers the best product available, but which also possesses a first class network that provides excellent data bases on its internal operations, on real and prospective customers, and on anything else that may be germane to its affairs – and which can act quickly on the information it has available.

David Harvey (1989b) conceives the sum of these processes as resulting in what he calls 'time-space compression' (p.284). This has been taking place over centuries, but since the early 1970s it has entered a particularly intense phase during which one-time limitations of space have been massively reduced (courtesy of information networks corporations can orchestrate their interests across huge distances) and the constraints of time have been eased (real-time trading is increasingly the norm in an age of global networks). It is certainly true that an important element of time-space compression has been the spread of rapid means of transport, notably air travel which, in the course of but a few decades, has shrunk the distance between continents dramatically. But even more important has been the establishment of complex and versatile information networks which enable the continuous and detailed management of dispersed affairs with relatively little concern for the restrictions of time. When one considers, say, the provision of perishable fruits and vegetables in a typical supermarket, food from around the world, made available the whole year round, one begins to appreciate what 'time-space compression' means for life in the late twentieth century. Much the same imagination can be applied to the manufacture and supply of microchips, fridges, clothes and even books.

These features each suggest a quality that is always highlighted in descriptions of post-Fordism – flexibility. This is posed as a distinct contrast with the circumstances which prevailed under Fordist regimes that were characterised as cumbersome, structured and standardised. Let us review some of the commonly considered aspects of flexibility and, as we do so, bear in mind that Fordist times were allegedly characterised by their opposites.

First, there is a new *flexibility of employees*. That is, the post-Fordist worker is one who neither holds to rigid job descriptions nor has the attitude that once equipped for an occupation he stays there for the rest of his working life. In contrast to the era of 'demarcation disputes' and 'once a fitter always a fitter', today we have adaptability as a central quality, with 'multi-skilling' the norm. Here the image is projected of 'lifetime training', of realisation that change is continuous in these 'new times', and that therefore employees must above all be 'flexible' (cf. McGregor and Sproull, 1992). Orientations to the job and to training are just one facet of this flexibility since there is also *wage flexibility* (a trend towards paying individuals for what they do rather than at an agreed union or national rate), *labour flexibility* (be prepared to change jobs every few years, to which end it is increasingly common to be employed on fixed-term contracts), and *time flexibility* (part-time employment is growing fast, as is 'flexi-time' and pressures to work shifts and, frequently, through the weekends).

Second, there is *flexibility of production*. Here the proposition is that Fordist methods are outdated by the spread, thanks to information networks, of more versatile and cost-effective production such as 'Just-in-Time' systems which wait until orders are taken before the factory manufactures, hence saving on warehousing and, of course, on unsold products. To function such systems must be flexible enough to respond with alacrity since, of course, customers will not wait long for the goods they have requested. Nonetheless, market competition puts a premium on such flexibility and impels corporations to invest in the information systems that can deliver it. Another form of flexible production is the vertical disintegration trend referred to above. It is evident that extensive use of subcontracts provides the corporation with the option of painlessly switching suppliers and products without the burden of offloading its own personnel.

Third, there is *flexibility of consumption*. Here the suggestion is that electronic technologies allow factories to offer more variety than was possible in the uniform Fordist period. Nowadays shorter runs are cost effective because computerisation provides the assembly line with unprecedented versatility. In addition, customers are turning against the uniformity of Fordist products, looking for *different* things which might express their own particular lifestyles and dispositions. Thanks to the information and communication infrastructure, goes the argument, customers' desires can at last be satisfied, with increasing amounts of customisation of production in the post-Fordist epoch.

These elements of flexibility, it ought to be understood, are in practice combined to a greater or lesser degree. In the archetypical post-Fordist organisation the customer's order is received, its particulars are routed to the factory where the plant is programmed to meet the individual specifications and a multi-skilled workforce sets to and manufactures what is required with adaptability and urgency. Note too that the entire process hinges, at each stage, on information processing, application and distribution. From the level of ordering through to that of supply a rapid, versatile and sophisticated information network is the *sine qua non* of everything.

It follows from these trends that we may observe in the post-Fordist era the *decline of mass production*. In place of huge and centralised plant what emerges are globally dispersed units employing in any one place only a few hundred people at the most, though world-wide the organising corporation is likely to have many more locations than before. In metropolitan centres opportunities for transnationals to reorganise internationally have exacerbated this trend, often leading to the movement of production to offshore and out of town locations, while occupations such as those in banking, insurance and business services have mushroomed (in Britain they doubled between 1971 and 1991, to almost one in eight of all occupations: OECD, 1993b: 456–457) since they offer crucial information services in key urban locations.

What this further signals is profound changes in the sort of jobs available in countries such as Britain. The male industrial worker is becoming outdated, to be replaced by part-time females on fixed-term contracts in the service sector. Manufacturing jobs have, since about 1970, been in 'steady and seemingly irreversible' (Pollard, 1983: 281) decline and it is especially women who have entered the 'flexible workforce' (Hakim, 1987). By 1991 little more than a quarter of all jobs were left in industry; services accounted for almost 70 per cent, and the majority of these are performed by women (OECD, 1993b: 456–457). In many organisations there appears to be a pattern of downsizing to a core group of permanent employees, and increased flexibility introduced by drawing on a large pool of peripheral labour. *Fortune* (24 January 1994) describes this as the 'contingency workforce' (those employed only when circumstances are favourable – and dropped as soon as they are not) which it estimates at 25 per cent of the American labour market. Within work, the emphasis is increasingly upon the versatile, information-oriented employee, at the upper levels those managerial groups whose numbers have burgeoned with restructuring and globalisation, but even lower down 'information jobs' are growing in the clerical, sales and secretarial realms.

The emergence of post-Fordism transforms geographical areas too, breaking up regions formerly distinctive in their work, class and political outlooks. Accompanying this is a shake-up of political and social attitudes. The mass industrial worker, his solidaristic unionism and his collectivist presumptions have little appeal to the post-Fordist citizen. Instead we have a revitalised enthusiasm for individualism and the 'magic of the market', which replaces the discredited planning of the post-war years. Kenneth Morgan (1990) goes so far as to argue that 'if there is one supreme casualty in British public life . . . it is the ethos of planning' (p.509), an ideology seemingly out of touch with the rapidity of change and *laissez-faire* operation of these 'new times'.

Nowadays indeed it seems that even the language of class has lost its salience. There is markedly less interest in class contours, conflicts and inequalities. It all seems, well, redolent of the 1960s, Alan Sillitoe, the dreary industrial North . . . . There is interest in an underclass, thought to inhabit the inner city ghettoes and isolated parts of the regions, but significantly it is considered a tiny group *detached* from the vast majority of society, separate and self-perpetuating which, if an irritant to law-abiding travellers, is distinctly apart from the bulk of the populace which is mortgage owning, self- and career-centred.

It is commonplace now to insist that the majority of the population is to be understood in terms of *different lifestyles*. In the post-Fordist regime class categorisations, and with them an associated common culture (the working-class male: work, community, club, mates, pigeons, football, horses, beer . . .) have given way to consideration of *differentiated* ways of life, to choices, options and customisation of production.

Some commentators insist that this results in the fragmentation of people's identities, in a loss of stability and satisfactions, while to others it is a democratising force which opens up new experiences and opportunities, stimulates the 'decentred' self and generates excitement. However, whatever differences of viewpoint here, the condition of post-Fordism is agreed upon: there is a new individualism around, an acknowledgement of variable lifestyles, and a recognition that class – it stands accused of being but a construction of the sociologist which is imposed on subjects of study – has lost force as a predictor of other dimensions of attitude and behaviour and as a basis for mobilising people on the political or industrial front.

We can appreciate here yet again how information and information circulation play an especially pertinent role in the post-Fordist regime. As Fordism is transformed from a production to a consumption-oriented system, not only is there a decline of the mass industrial worker, but also there emerges a more individualist and consumption-centred person. Information necessarily takes on a greater role in his or her life, first because consumers must find out about what is available to consume and, second, because in the individualised present they are eager to make statements about themselves through their consumption. Both factors promote information, the former because it concerns advertising and promotion of goods and services, the latter because it involves the symbolic dimensions of consumption, people using objects and relationships to make statements about themselves, thereby generating more information.

OBJECTIONS

For a good deal of the 1980s Fordist/post-Fordist theorisations attracted much attention in intellectual circles. For some interest, came from the search to explain the demonstrable inability of the Left in Britain to win electoral support, voters recurrently (in 1979, 1983, 1987 and 1992) unwilling to endorse collectivist appeals and antipathetic to the dated image of the Labour Party. There just had to be some reason for this failure; after all, the people had frequently supported Labour between 1945 and the 1970s, so what had changed? More generally, there was widespread awareness of rapid transformations taking place which convinced many commentators that something radically different was coming into being. Not surprisingly perhaps, a great deal of writing was produced which highlighted the 'New Times' (1988).

Unfortunately, however, it is precisely this emphasis on radically 'new times' conjured by the concept *post*-Fordism that causes the most difficulty. The suggestion is that society has undergone deep, systemic, transformation. And, indeed, what else is one to conclude when post-Fordism's characteristics are presented as so markedly *different* from what has gone before? (cf. Hall and Jacques, 1989).

It is because of this that one may note an ironic congruence between post-Fordism and the conservative post-industrial society theory of Daniel Bell that we encountered in Chapter 3, there being a shared concern to distinguish the present from the recent past, to depict a new age coming into being, albeit that the conceptions have significantly different intellectual traditions. In fact, Krishan Kumar (1992) goes so far as to identify post-Fordism as a 'version of post-industrial theory' (p.47), one which concerns itself with remarkably similar themes and trends.

Against this it is salutary to be reminded that, to the extent that private property, market criteria, and corporate priorities are hegemonic, and these are acknowledged to be such at least in Regulation School versions of post-Fordism, then a very familiar form of capitalism still pertains. Hence it might be suggested that the term neo-Fordism, with its strong evocation of the primacy of *continuities* over change, is more appropriate. Put in this way, the suggestion is that neo-Fordism is an endeavour to rebuild and strengthen capitalism rather than to suggest its supersession.

Most objections, at least to strong versions of the theory, centre of the conception's tendency to emphasise change over continuity. This leads adherents perhaps too readily to endorse a binary opposition (Fordism *or* post-Fordism) which oversimplifies historical processes and underestimates the uninterrupted presence of capitalist relations through time. I limit myself here to signalling some of the more telling criticisms of the theory (Sayer, 1989).

The depiction of Fordism suggests an equilibrium that was far from existing between 1945 and 1973. For example, in Britain between 1950 and the mid-1970s one-third of farm workers' jobs were lost (Newby, 1979: 81), a striking feature of the agricultural landscape, but one which brought forth no theories of profound social change.

Indeed, when one comes across post-Fordists insisting that, for example, class politics is outmoded because the working class (taken to be manual workers) is disappearing, it is as well to remember that the *industrial* working class has always been in a minority in all countries *except* Britain (and even there it only just constituted a majority for a short period), and that manual work for much of modern history has been undertaken very largely by agricultural labourers. In Britain, for instance, farm workers accounted for 25 per cent of the occupied population in the mid-nineteenth century, more than the total of those engaged in mining, transport, building and engineering (Hobsbawm, 1968: 283, 279). Agriculture's continual decline since then (now it accounts for less than 3 per cent of total employment) highlights the fact that the working class has a long history of recomposition (Miliband, 1985) with certain occupations growing and others in decline.

This being so, we might then also be sceptical of those commentators who conclude that a steady growth of white-collar work announces the end of the working class. This very much depends upon one's definitional criteria. The expanding army of non-manual employees certainly does have particular characteristics, but it may be premature to assume that they are more decisively differentiated from the factory worker today than was the engineering tradesman from the agricultural labourer at the turn of the century. Moreover, recollecting these sort of divisions within manual occupations, we might usefully reflect on the fact that there has never been the period of working-class homogeneity suggested by the Fordist typology.

It is as well to hold in mind that the equation of manual work with the working class, and this, with a homogeneity of outlook, is very much a construction of intellectuals. It may imply a confluence that in reality is absent, just as it may suggest an unbridgeable gulf separating the working class from white-collar (and thereby middle-class?) work. Finally, while we ponder these problems, we might also remember that manual work has far from disappeared in the 'post-Fordist' 1990s – in Britain today it is still little more than a few points off from amounting to half the total workforce.

Post-Fordism makes a good deal of the decline of work in factories and the shift to service occupations such as in finance and leisure. This is undeniably empirically true, but it is hard to contend that it marks a really profound change. On the contrary, the spread of many services is to be explained by divisions of labour introduced to make more effective capitalist activity.

The post-Fordist emphasis on consumption, to which I return in the next chapter, has met with many objections. Prominent amongst these are the following.

1 The observation that consumption has a long history casts some doubt on post-Fordist theorists' portrayal of its novelty. Consumption has been a concern since at least the latter part of the eighteenth century when industrial techniques began to make consumer goods available on a wide scale (McKendrick *et al.*, 1982). Seen from a long-term perspective recent developments may indicate an acceleration of trends, but scarcely a seismic change from 'production to consumption'.

2 The argument that consumption is characterised by increased individuation amongst people (the stress on difference) and a capacity amongst manufacturers to supply correspondingly customised products is questionable, particularly in its contention that this signals a marked difference from the Fordist era of 'mass consumption' and 'mass production'.

A number of objections are made to this, chief amongst which is that mass consumption and mass production continue unabated in the 1990s. While during the 1960s this came in the form of television and automobiles, in the 1990s it is still cars, but also video recorders, compact disc players, home computers and dish-washers, fitted kitchens, flat-pack furniture and the like, which represent the latest generation of mass-produced consumer goods.

Further, the assertion of post-Fordists that mass consumption is antipathetic to individualism (the image of the dull and dreary 1950s is always evoked) is dubious, not least because it is perfectly possible to employ mass-produced goods in ways which reinforce one's sense of individuality. For example, one may select from a variety of mass produced clothes combinations which when mixed are unusual and suggest individuality. Indeed, modularisation of consumer products, a conscious strategy of corporate suppliers, is an endeavour to manage consumers' desire for choice *within* a framework of continuing mass manufacture.

Observing that mass production remains preponderant leads one to consideration of those responsible for organisation of the corporate sector. Here one of the recurrent themes of post-Fordist theory is that in the present era the emphasis on flexibility provides opportunities for small, fast-paced and innovative organisations to enter markets and best their bigger competitors because they can be more responsive to consumer needs.

Against this it should be reiterated that the history of the last thirty or so years has been one of the unabated expansion and aggrandisement of long-established corporations. Amongst the major characteristics of globalisation has been the continued pre-eminence of transnational corporations which, wherever they operate, account for the lion's share of the market. *Any* examination of the leading sectors of *any* market of economic significance will bear that out. Indeed, what is particularly impressive is the way in which so many corporate leaders of the early decades of the twentieth century continue to retain their prominent positions at the forefront of today's globalised economy – for instance Ford, General Electric, Shell Oil, Siemens, Proctor and Gamble, Daimler-Benz, Coca-Cola, Kellogg, IBM, ICI, Kodak, Philips, General Motors and Fiat. What the evidence indicates here is that there are fundamental continuities (odd name changes and amalgamations apart) in post-war (and even pre-war) history, something which must make one hesitant to announce any 'post' developments.

There is little evidence to suggest that these industrial titans cannot respond to, or even create, consumer diversity in their production activities. Adoption of new technologies, allied to more versatile marketing, means that TNCs are 'quite adept at mass producing variety' (Curry, 1993: 110). One of the false premisses of much post-Fordist theory is that global corporations are somehow incapable of responding with alacrity to local and particular needs. But there is no logical incompatibility between global reach and local responsiveness. Indeed, astute marketers, armed with appropriate information bases and networks, are well able to *target* customers distributed around the globe and organise production appropriately. Thereby globalism and local responsiveness can be harmonised in what Kevin Robins (1991b) calls the 'flexible transnational' (p.27) corporation.

FLEXIBLE SPECIALISATION

Such criticisms of post-Fordist conceptions do carry weight, but they can always be responded to, at least by Regulation School-influenced theorists, by the insistence that what is being considered is not an entirely new system, but rather a mutation of capitalist regimes of accumulation. One can complain of ambiguity and uncertainty in their analyses – how much is continuity, how much is change? – but because most authors start their accounts from a broadly Marxian perspective which is interested in the dynamics of capitalism there always remains the defence, to the charge that capitalist relations continue, that all that is being identified is another mode of capitalist enterprise.

However, there is at least one influential school of thought which, starting from a more focused position, presents a variant of post-Fordism that does suggest a more decisive break with the past. The writing of Michael Piore and Charles Sabel (1984), centring on work, suggests that the spread of flexible specialisation/production offers the prospect of widespread improvement in ways of life. Because this theorisation places particular emphasis on the role of information/knowledge in post-Fordist work situations, it merits here separate review from the more general Regulation School theory.

The argument is that during the era of Fordism, when mass production predominated, large volume manufacture of standardised products demanded specialisation of machinery and a congruent specialisation of labour which was, unavoidably, characterised by low levels of skill. Conjure up the image of the assembly line in the large factory and one can readily picture this scene. It was one in which Taylorist techniques (rigid time and motion, hierarchical supervision, restriction of operatives to narrowly conceived routines designed by management) were the norm and semi-skilled and unskilled labour the typical requirements.

Piore and Sabel contend that 'we are living through a second industrial divide', comparable to the first which brought about mass production in the late nineteenth century. The most recent heralds 'flexible specialisation', a radical break with the repetitious and low-skilled labour of Fordism, one which will increase the skills of employees and allow greater variety in the production of goods. This flexibility is the keynote of the new age, a chord already struck in the Italian region of Emilia-Romagna (Sabel, 1982), and one which portends an end to stultifying labour and a return to craft-like methods of production in small co-operative enterprises that can respond rapidly to shifting market opportunities.

Three main reasons are adduced to explain the emergence of flexible specialisation. First, it is suggested that labour unrest during the 1960s and the early 1970s encouraged corporations to decentralise their activities by increasing the amount of subcontracting they used and/or divesting themselves of in-house production facilities. This stimulated the spread of small, technically sophisticated firms, themselves often established by those displaced in consequence of the restructuring strategies of large firms, but eager for work, possessing high skills, and adaptable. Second, changes in market demand have become evident, with a marked differentiation in consumer tastes. This provided opportunities for low volume and high-quality market niches to which flexible specialisation was well adapted. Third, new technologies enabled small firms to produce competitively because the advantages of economies of scale were reduced as skilled outfits began to maximise their versatility thanks to the flexibility of modern computers. More than this though, the new technologies, being extraordinarily malleable through appropriate programming, at once increase the competitive edge of the fast-footed small firm and upgrade existing skills because they 'restore human control over the production process' (Piore and Sabel, 1984: 261).

This is a simplification of flexible specialisation theory, but for my purposes it is necessary only to make two points. The first concerns the quite extraordinary diversity of opinion which endorses the notion. In what appears to be a generalised reaction against Harry Braverman's (1974) once popular contention that capitalist advance results in the progressive deskilling of labour (cf. Penn, 1990), a host of thinkers now announce flexible specialisation as the coming of an age which may upskill employees. In the UK these thinkers range from economist John Atkinson (1984) whose early studies of the 'flexible firm' struck a chord with political and business leaders who pressed for a flexible workforce as a response to competitive threats and recession (Atkinson and Meager, 1986), to Paul Hirst and Jonathan Zeitlin (1991) emerging from a Marxian tradition to contend that flexible specialisation may be formed anywhere where favourable patterns of 'co-operation and co-ordination' exist that supply the necessary 'irreducible minimum of trust' between workforce and employers (p.447) to make it happen. Across the Atlantic there is a correspondingly wide range of exponents, from radical critics like Fred

Block (1990) who see 'postindustrial possibilities' bringing 'higher skill levels' (p.103), to Soshana Zuboff (1988) of the Harvard Business School who discerns the prospect of 'a profound reskilling' (p.57) in recent developments. 'Flexible specialisation' now passes almost for conventional wisdom across a wide spectrum of intellectual opinion.

The second point is that *information* is regarded as having an axial role to play in flexible specialisation, in several ways. One is that, concentrating on production work as many of these writers do, information technologies are arguably the major facilitator and expression of flexibility. The new technologies are 'intelligent', their distinguishing feature being that they incorporate considerable quantities and complexities of information. The programs which guide them are their fundamental constituents rather than any specific function they may perform. It is these information inputs which determine their degrees of flexibility, enabling, for example, cost-effective small batch production runs, customisation of products, and rapid changes in manufacturing procedures. Furthermore, it is this information element which provides flexibility in the labour process itself, since to perform the operatives must, of course, be 'multi-skilled' and adaptable, hence more flexible (which in itself promotes the role of information). Where once upon a time employees learned a set of tasks 'for life', in the age of information technology they must be ready to update their skills as quickly as new technologies are introduced (or even reprogrammed). Such 'skill breadth' (Block, 1990: 96) means employees have to be trained and retrained as a matter of routine, a pre-eminently informational task.

Another way in which information is crucial also stems from this increased reliance on programmable technologies. The very fact that the machinery of production is so sophisticated requires that workers possess information/knowledge of the system as a whole in order to cope with the inevitable hiccups that come with its operation. Thus not only does information technology stimulate regular retraining, but it also demands that the employees become knowledgeable about its inner workings. In this way production workers become in effect information employees. In the terminology of Larry Hirschhorn (1984) these are 'postindustrial workers' who 'must be able to survey and understand the entire production process so that they are ready to respond to the unpredictable mishap' (p.2). Information technologies on the shop floor are a 'postindustrial technology' (p.15) which takes away many of the physical demands and tedium of assembly work, but also requires 'a growing mobilization and watchfulness that arises from the imperfections, the discontinuities of cybernetic technology'. Therefore, 'learning must be instituted in order to prepare workers for intervening in moments of unexpected systems failure', something which requires comprehension of the overall system. We may foresee 'the worker moving from being the controlled element in the production process to operating the controls to controlling the controls' (pp.72–73).

More than this, flexible specialisation also encourages employee participation in the design of work. That is, computerisation of production provides 'cybernetic feedback' (Hirschhorn, 1984: 40) to the operative which enables him or her to act by reprogramming the system in appropriate ways. Here we have the worker depicted as informationally sensitive, made aware by advanced technologies of what is happening throughout the production process, and able to respond intelligently to improve that overall system. It is this to which Soshana Zuboff (1988) refers as the *reflexivity* that comes from working with IT, an 'informating' (p.10) process which she believes generates 'intellective skill'.

Scott Lash and John Urry (1994) take this reflexivity element to greater heights, *en route* relegating the emphasis on information technologies in favour of information itself, while also taking aboard concern for areas of work other than those involved with production. In their view we inhabit an era of 'reflexive accumulation' where economic activity is premissed on employees (and employers) being increasingly self-monitoring, able to respond to consumer needs, market outlets and rapid technical innovation, with maximum speed and efficacy. In such circumstances information occupies centre stage since it is this which is the constituent of the vital reflexive process that guides everything and which is a matter of continuous decision making and amendment on the basis of ongoing monitoring of processes, products and outlets.

In addition, production of things has become infused with symbols in so far as *design* elements have become central to much manufacture while, simultaneously, there has been an explosive growth of work which is primarily and pre-eminently symbolic. These changes are manifest in the motor industry (where, after all, a great deal of innovation is a question of design rather than narrowly conceived technical refinement), but how much more have they penetrated the music business, television production and publishing, fast-expanding cultural industries where information soaks into every aspect of work (ibid.: 220–222).

The contention here is that work increasingly features 'design intensity' as its informational dimensions move to the fore, whether it is in the manufacture of 'stylish' clothing and furniture or whether it is in the area of tourism and entertainment. Further, against the perception that work is largely a matter of routinised factory production, Lash and Urry emphasise ways in which even goods production has been influenced by wider developments which impel products to incorporate cultural motifs (they have been 'aestheticised') and which intrude into work relations so as to inculcate a 'university'-like ethos in pioneering areas such as the IT industry. Given such trends, Lash and Urry believe that work can take one of two forms; either innovation can be devolved to the shop floor and operatives allowed a larger role in the process (in the manner of Hirschhorn), or it can bypass the shop floor altogether, with its functions taken over by 'professional-managerial workers' (p.122) such as are found already in the high tech and advanced producer and consumer services. Either way, the prospect is for high-skilled work in the era of 'flexible specialisation'.

These ideas of flexible specialisation, with the suggestion of work being information-intensive and of higher skill levels than hitherto, are understandably appealing. Still more attractive, one can recognise the professionalised employee in the cultural industries, eagerly on the lookout for new 'ideas' or 'styles' to take up and explore, dealing all the time with information in a reflexive manner while searching out market niches by constantly innovating. The writer of self-help books, the travel guide, the producer contracted to Channel 4, perhaps even the university professor are all of this type.[7]

However, theories of 'flexible specialisation' have had to encounter a great deal of hard-headed criticism. Prominent amongst these are the following.

- Some of the advocates show a strong trace of technological determinism (C. Smith, 1989). Those such as Hirschhorn (1984) who place emphasis on the cybernetic capabilities of computers fall too easily into a tradition which presumes that advanced technologies bring with them advanced skill requirements. From his perspective 'industrial technology' is 'transcultural', unavoidably 'shap[ing] social life in the same mold everywhere' (p.15), only to be broken (and liberated) by 'postindustrial technology' (*sic*) which brings flexibility. This has been vigorously challenged empirically (Wilkinson, 1983; Shaiken, 1985) and theoretically (Webster and Robins, 1986, pp. 49–73).
- 'Flexible specialisation' is presented as the opposite of mass production and with this in some way contrary to the continuing dominance of large corporate organisations. However, it is doubtful whether this is the case, for several reasons. One, which has already been reviewed, is that it underestimates the flexibilities of giant corporations that are well able to introduce into their affairs new modes of working, new technologies that enhance versatility, and modular products that allow for significant product differentiation while continuing mass production practices.

  As Charles Sabel (1982) concedes, 'existing Fordist firms may be able to meet the changing demand without sacrificing their fundamental operating principles' (p.194). Case studies of large motor manufacturers indicate this possibility. Nissan, for example, established new and flexible production plant in Sunderland, but continued relations which entail close control over a subordinated labour force (Garrahan and Stewart, 1992).

  Another objection is that, in spite of undoubted examples of flexible specialisation that may be found, mass production remains dominant throughout the advanced economies. Thus any suggestion of a marked change is empirically false (Williams *et al.*, 1987). Still another objection is that there is little new about flexibility since it has been a feature of capitalist enterprise since its origination (Pollert, 1988: 45–46). The nineteenth century is replete with instances of specialist enterprises to meet market segments, but no one has ever felt compelled to present say the rag trade or toy makers (cf. Mayhew, 1971) as illustrative of flexible specialisation.

  Connectedly, while enthusiasts present flexible specialisation in positive terms, it can be interpreted as the re-emergence of what others have termed 'segmented labour'. That is, while there may indeed be a core of confident, skilled and versatile employees, there are also identifiable much more vulnerable (and hence flexible) 'peripheral' people working part-time, casually or on short-term contracts (Gordon *et al.*, 1982). Arguably these 'peripheral' groups have expanded in recent years, though there is some doubt about quite how much this has happened and certainly they have long been a feature of capitalist enterprise.

- Perhaps the sharpest attack has come from Anna Pollert (1988, 1990) who criticises the vagueness and catch-all character of 'flexibility' which, when broken down into more testable elements (flexibility of employment, of skill, of time, of production), loses much of its force and originality (though always it appears that it is the employee who is called upon to be flexible: cf. Clarke, 1990a, 1990b).

CONCLUSION

This chapter has undertaken a review of claims that there has been a transition from a Fordist to a post-Fordist regime of accumulation and the related argument that mass production has given way to flexible specialisation. It is irritatingly difficult to sum up the state of the debate since a good deal of the argument is ambiguous and uncertain, unwilling to state directly whether we are supposed to have experienced a systemic change or whether what has emerged is more a continuation of established capitalist relations.

What is clear is that we ought to be sceptical of suggestions that we have undergone a sea change in social relationships. Features of capitalist continuity are too insistently evident for this: the primacy of market criteria, commodity production, wage labour, private ownership and corporate organisation continues, establishing links with even the distant past. Nonetheless, it is surely indisputable that, over the post-war period, we can observe some significant shifts in orientation, some novel forms of work organisation, some changes in occupational patterns and the like. We should not make the mistake of going beyond acknowledgement of these changes to the contention that we have witnessed a system break of a kind comparable with, say, slavery's supersession by feudalism or, more recently and certainly more profound than any Fordism to post-Fordism transition, the collapse of Communist regimes and the attempts to replace these with market-based systems.

This qualification aside, I believe that several major changes in post-war capitalist organisation may be registered:

- The deep recession that hit capitalist societies in the 1970s impelled a restructuring of relationships which inevitably resulted in upheaval and instability.
- The process of globalisation, in its diverse aspects, continued and accelerated, making it untenable for corporations to continue as before, and presented them with challenges and opportunities that had to be met.
- Throughout the period transnational corporations expanded in size, scope and reach, in ways without historical precedent; this made them the major players in the global economy.

Combined, these developments precipitated major changes in capitalist activity, not least an acceleration of change itself, something which encouraged more flexible strategies of production, marketing and, to some degree at least, consumption. And absolutely axial to these developments, and to the handling of change itself, was information, from the level of the factory and office floor to world-wide corporate operations.

Information may not have brought about these changes, but today it indisputably plays a more integral role in the maintenance and adaptability of capitalist interests and activities. By way of a conclusion, let us signal some of the crucial contributions that information makes.

- Information flows are a requisite of a globalised economy, particularly those financial and service networks which tie together and support dispersed activities.
- Information is central to the management and control of transnational corporations, both within and without their organisations.
- Information is crucial to the emerging phenomenon of global localism, whereby international and local issues and interests are connected and managed.
- Information now plays a more integral part in work practices, at once because computerisation has pervasive effects and also because there has been a noticeable increase in the information intensity of many occupations.

8

INFORMATION AND POSTMODERNISM

Postmodernism has been much in evidence in a wide variety of academic and media realms since the early 1980s. The word crops up in just about every university discipline from art history to accountancy, while outside academe it is very commonly used to describe anything from architectural styles to rock music. Despite its widespread currency it is a difficult subject with which to come to grips. Its meanings are variable and, where they are not, are just as likely to be obscure.

Nonetheless, what I want to undertake in this chapter is a discussion of the relations between information and postmodernism. I shall focus on a number of key figures who pay particular attention to the informational aspects of postmodernism. Preliminary to this, however, I shall attempt to define postmodernism in reasonably straightforward terms – no easy task in itself, since, as we shall see, it is hard to identify the essence of something which denies the reality of essences! Finally, I shall comment on discussions of postmodernism which present it as the outcome of social and economic changes. Here thinkers identify postmodernism as a *condition* which is consequent on changes that are open to examination by established social analysis.

It needs to be made clear right away that these scholars who conceive of a *postmodern condition* are significantly different from *postmodern thinkers* who reject the entire approach of those who endeavour to explain the present using the conventions of established social science. That is, we may distinguish the position of those who argue that we may conceive of a *reality* of postmodernism, from that of postmodern thinkers who argue that, while we do indeed inhabit a world that is different – and hence postmodern – from anything that has gone before, this very difference throws into doubt the validity of orthodox tenets of social explanation. This somewhat philosophical point may not appear important here but, when we come to analysis of postmodern scholars, it will become evident that the openness to examination of their descriptions of contemporary society by orthodox – one might say *modern* – social science significantly influences one's willingness to fully endorse their points of view.

POSTMODERNISM

Postmodernism is at once an intellectual movement and something that each of us encounters in our everyday lives when we watch television, dress to go out, or listen to music. What brings together the different dimensions is a *rejection of modernist* ways of seeing. It is, of course, an enormous claim, to announce that postmodernism is a break with ways of thinking and acting that arguably have been supreme for several centuries.

Much of the claim depends, of course, on what is meant by the terms *postmodern* and *modern*. Unfortunately, many of the relevant thinkers either do not bother to state precisely what they mean by these words or they concentrate only upon certain features of what they take them to be. That said, within the social sciences *modernity* is generally understood to identify a cluster of changes – in science, industry and ways of thought that we usually refer to as the rise of the Enlightenment – that brought about the end of feudal and agricultural societies in Europe and that has made its influence felt pretty well everywhere in the world. *Postmodernity* announces a fracture with all of this.

Some commentators have argued that *postmodernism* ought to be considered more as a matter of culture than the above, in that its concerns are chiefly about art, aesthetics, music, architecture, movies and so forth (cf. Lash, 1990). In these cases the couplet modernism/postmodernism is less overarching than the distinction between modernity and postmodernity. Moreover, if we restrict ourselves to this cultural arena, then there is less of a willingness to announce a break with modernism since, of course, Modernism – with a capital M – refers to movements of the late nineteenth and early twentieth centuries – Impressionism, Dadaism, Surrealism, Atonalism and so on – which themselves stood in opposition to classical culture.

Modernism refers to a range of movements in painting, literature and music which are distinguished from classical forms in that the latter were committed to producing culture which was determinedly representational. Think, for instance, of the 'great tradition' of nineteenth-century realist English novelists, dedicated to telling a story which was clear and evocative, 'like real life', or consider so much painting of this era which was portraiture, aiming to produce accurate likenesses of subjects. Modernist writers such as Joyce and painters such as Picasso decisively broke with these predecessors.

With regard to postmodernism there are at least two difficulties to be encountered here. The first concerns the matter of chronology. Modernity commences around the mid-seventeenth century in Europe, while modernism is very much more recent and that which it opposed – classical culture – was itself a product of the period of modernity. With modernity predating modernism, plus modernity being a concept that embraces an extraordinary range of changes from factory production to ways of thought, the question of modernism's relationship to modernity is problematical and is a source of conceptual confusion. Is modernism/postmodernism a subsidiary element of the modernity/postmodernity divide?

The second problem is that postmodernism does not announce a decided break with modernist cultural principles, since at the core of postmodernism is a similar refusal of representational culture.

Were one to restrict oneself to a cultural notion of postmodernism it would be possible to argue that the implications of the 'post' designation are fairly minor, restricted to relatively few areas of life and building upon the premisses of modernism. Such a conception is much less grand and ambitious than the announcement of postmodernity which rejects modernity *tout court*.

Distinguishing modernity/postmodernity and modernism/postmodernism might appear useful in that it could allow us to better understand the orientation of particular contributions to debates. Unfortunately, however, it is of little practical help because most of the major contributors to the debate about postmodernism, while they do indeed focus upon cultural phenomena, by no means restrict themselves to that. Quite the contrary, since for them the cultural is conceived to be of very much greater significance now than ever before, they comfortably move on to argue that postmodernism is a break with modernity itself. Hence very quickly postmodern thinkers move on from discussions of fashions and architecture to a critique of all expressions of modernity in so far as they claim to represent some 'reality' behind their symbolic form. For example, postmodern thinkers reject the pretensions of television news to 'tell it like it is', just as quickly as they reject the pretensions of social science to provide accurate information about the ways in which people behave. From the cultural realm wherein it punctures claims to represent a reality in symbolic forms to the presumptions of thinkers to discern the major dimensions of change, postmodernism insists on the radical disjuncture of the present with three centuries and more of thought.

Therefore we need not be overconcerned about limiting postmodernism to the realm of culture, since its practitioners themselves show no similar compunction. Quite the reverse: postmodernism as an intellectual movement and as a phenomenon we meet in everyday life is announced as something radically new, a fracture with modernity itself. Let us say something more about it.

INTELLECTUAL CHARACTERISTICS OF POSTMODERNISM

Seen as an intellectual phenomenon, postmodern scholarship's major characteristic is its opposition to what we may call the Enlightenment tradition of thought which searches to identify the *rationalities* underlying social development or personal behaviour. Postmodernism, influenced heavily by Friedrich Nietzsche (1844–1900), is deeply sceptical of accounts of the development of the world which claim to discern its growth, say, in terms of fundamental processes of 'modernisation', and it is equally hostile towards explanations of personal behaviour that claim to be able to identify, say, the foundational causes of human 'motivation'.

Postmodernism is thoroughly opposed to each and every attempt to account for the world in these and similar ways, all of which seek to pin point rationalities which govern change and behaviour. The presumption of Enlightenment thinkers that they may identify the underlying rationalities of action and change (which may well go unperceived by those living through such changes or acting in particular ways) is a focus of dissent from postmodernists.

This is generally voiced in terms of hostility towards what postmodernists call *totalising* explanations or, to adopt the language of Jean-François Lyotard, 'grand narratives'. From this perspective all accounts of the making of the modern world, radical or conservative, that claim to perceive the mainsprings of development in such things as the 'growth of civilisation', the 'dynamics of capitalism', and the 'forces of evolution' are to be resisted. It is undeniable that these and similar analyses are endeavouring to highlight the major trends and themes – the main rationalities – of human development. Postmodern thinkers resist them on several related grounds.

The first principle of resistance is that these accounts are the construct of the theorist rather than accurate studies of historical processes. Here scholars who adopt the Enlightenment presumption that the world is knowable in a reliable and impartial way are challenged. Their identification of rationalities stands accused of being an expression of their own perception rather than a description of the operation of real history. This criticism is a very familiar one and it is axiomatic to postmodern thought. In brief, it is the charge that all external claims for the validity of knowledge are undermined because scholars cannot but interpret what they see and, in interpreting, they are unavoidably involved in *constructing* knowledge.

The second and third points of resistance show that this is not a trivial philosophical objection. This is because the 'grand narratives' which lay claim to demonstrate the 'truth' about development reveal their own partialities in so far as the logical outcome of their studies is recommendation, if often implicit, of particular directions social change ought, or is likely, to take. Moreover, not only is the accusation made that totalising accounts of social change are but a prelude to planning and organising the present and future, the charge is also brought that these have been thoroughly discredited by the course of twentieth-century history.

For example, studies of social change which suggest that the most telling forces of development are the search for maximum return for minimum investment are, clearly, trying to identify the predominant rationality to have governed change. It matters not that for some historical periods and in some societies this rationality has not been followed, since such 'irrationalities' are usually regarded as aberrations from a decisive historical directionality.

Reflection on this approach to history does reveal that its claim to chart the course of the past tends to carry with it implications for future and present-day policy. It implies that the rationality of 'more for less' will continue to prevail and, frequently if not always, that planners ought to take responsibility for facilitating or manipulating events to keep things on track. This indeed has been an important consideration for many development theorists who have sought to influence policies towards the Third World on the basis of having discerned the successful rationality underpinning Western economic growth.

The accusation that these analysts who claim they are able to highlight the driving forces of change are partial finds supportive evidence in the increasing frequency with which their scholarship and the policies which draw upon them are discredited: by, for example, arguments that they disadvantage the 'underdeveloped' world, or that the 'more for less' rationality, due to its anti-ecological bias, is threatening to the survival of human and animal species on 'planet earth', or that the 'green revolution' which promised agricultural bounty by appliance of modern science has led to social dislocation, unemployment of displaced farm workers, and dependence on far-away markets.

A still more frequently considered example of the failure of 'grand narratives' is that of Marxism. Reflect that it has claimed to identify the mainsprings of historical change in the course of the 'class struggle' and 'capitalist accumulation'. In identifying the rationalities that have governed change it is evident that Marxist thinkers see these as being ultimately supplanted by a higher rationality. Their advocacy, which gains support from their historical studies, was that a new form of society would be established that could take advantage of, and overcome shortcomings in, capitalist regimes. However, Marxist claims to reveal the true history of social change are, in the aftermath of the disintegration of Soviet Communism, thoroughly discredited. Today Marxism is increasingly regarded as the construct of those with particular dispositions, a 'language' which allowed people to present a particular way of seeing the world.

To postmodernists such as Lyotard recent history has fatally undermined not just 'grand narratives', but all Enlightenment aspirations. Fascism, Communism, the Holocaust, super-sophisticated military technologies, Chernobyl, AIDS, an epidemic of heart disease, environmentally-induced cancers . . . all these are the perversions of Enlightenment, outcomes of 'narratives' of the past which insisted that it was possible to highlight the rationalities of change, whether in terms of 'nationalism', 'class struggle', 'racial purity' or 'scientific and technological progress'. In view of such outcomes the postmodernist urges 'a war on totality' (Lyotard, 1984: 81), an abandonment of accounts of the world which presume to see the 'true' motor(s) of history. All pretension to discern the 'truth' of historical change 'has lost its credibility . . . regardless of whether it is a speculative narrative or a narrative of emancipation' (ibid.: 37).

It follows from this that postmodern thought is characteristically suspicious of claims, from whatever quarter, to be able to identify 'truth'. Given the manifest failures of earlier 'grand narratives', given that each has demonstrably been a construct however much scholars have proclaimed their objectivity, then postmodernism rejects them all by endorsing a principle of relativism, by insisting that, where there is no 'truth' there can only be versions of 'truth'. As Michel Foucault (1980) put it, postmodernists perceive that 'each society has its regime of truth, its 'general politics' of truth: that is, the types of discourse which it accepts and makes function as true' (pp.131–132). In such circumstances postmodern thinkers perceive themselves to be throwing off the strait-jacket of Enlightenment searches for 'truth', emphasising instead the liberating implications of *differences* of analysis, explanation and interpretation.

SOCIAL CHARACTERISTICS OF POSTMODERNISM

In the social realm postmodernism's intellectual critique is taken up, restated and extended. Here we encounter not just postmodern thinkers, but also the circumstances which are supposed to characterise postmodern life. To appreciate the postmodern condition we do not have to endorse the postmodern critique of Enlightenment thought, though it will be obvious that, if we are indeed entering a postmodern world, then its intellectual observations will find an echo in the social realm. Moreover, since all readers of this book inhabit this postmodern culture they will want to test the following descriptions against their own experiences and perceptions.

As with the intellectual attack, a starting point for postmodernism in the social realm is hostility towards what may be (loosely) called modernist principles and practices (Kroker and Cook, 1986). Modernism here is a catch-all term, one which captures things such as planning, organisation and functionality. A recurrent theme is opposition to anything which smacks of arrangements ordered by groups – planners, bureaucrats, politicians – who claim an authority (of expertise, of higher knowledge, of 'truth') to impose their favoured 'rationalities' on others. For example, designers who presume to be able to identify the 'really' fashionable, to set standards for the rest of us of how we ought to dress and present ourselves, find their privileged status challenged by postmodern culture. Again, functionality is resisted on the grounds that the 'most efficient' way of building houses reflects, not some disembodied 'rationality' of the technically expert architect or town planner, but an attempt by presumptuous professionals to impose their values on other people.

What will be obvious here is that the postmodern mood is quizzical of judgements from on high. To this extent it contains a strong streak of, as it were, democratic impudence, something manifested in ready rejection of those who would define standards for the rest of us. Of particular note here is the antipathy postmodernism expresses towards received judgements of 'good taste' or the 'great tradition' in aesthetics. Against those who presume, on grounds that they are expert or perceptive enough to have access to 'true judgement', to rank, say, literature, the postmodern sensibility parades 'each to his own taste': 'If Jeffrey Archer is your bag, then who are these literature professors to tell you what is better?'

Those who set standards in the past are routinely decried. Thus Leavis (1977) might confidently assert that his evaluation of 'the great tradition' came from an especially close reading of the English novel, but the postmodernist readily enough demonstrates that the literary critics make a living out of their criticism, their writings bringing them career advancement and prestige (hence they are scarcely disinterested seekers after truth). Moreover, it is an easy task to reveal that the critics' valuations rest heavily on particular assumptions, educational background and class preferences. In short, the partialities of critics are exposed and thereby the legitimacy of their claims to impose their judgements on the rest of us taken away.

Unmasking the pretensions of 'true' thinkers, postmodern culture testifies to aesthetic relativism – in each and every realm of life *difference* is to be encouraged. This principle applies everywhere (Twitchell, 1992): in music ('Who is to say that Mozart is superior to Van Morrison?'), in clothing ('Jaeger doesn't look any better than Next, it just costs more'), as well as in the live arts ('Why should Shakespeare be privileged above Andrew Lloyd Webber?'). This has a liberatory quality since at postmodernism's centre is refusal of the 'tyranny' of all who set the 'right' standards of living one's life. Against these postmodern culture thrives on variety, on the carnivalesque, on an infinity of differences. For example in housing the Wimpey estate and the high-density tower block designed by those who presumed to know what was 'best for people' and/or 'what people want' are resisted, in their place the climate of opinion becomes one which tolerates individuating one's home, subverting the architect's plans by adding a bit here, knocking a wall down there, incorporating bits and pieces of whatever one pleases and let those who say it is in poor taste go hang.

At back of this impulse is, of course, the refusal of the modernist search for 'truth'. Postmodernism resists it, on the one hand, because the definers of 'truth' can frequently be shown to be less than ingenuous about their motivations and, anyway, there is so much disagreement amongst the 'experts' themselves that no one believes there is any single and incontestable 'truth' to discover any more, and, on the other hand, because it is evident to all of us that definitions of 'truth' easily turn into tyrannies. To be sure, nothing like the Communist regimes which ordered people's lives because the Party best knew the 'objective realities of the situation', but still each of us will have experienced the imposition of others' judgements on ourselves. At school we will have to have read Dickens and Hardy because definers of 'literary standards' deemed them to be worthy of inclusion on the curriculum (while ruling out science fiction, romance and westerns). Again, everyone in Britain will have experienced BBC television as that which cultural custodians had thought worthy of production (lots of news and current affairs, the classic serials, 'good' drama, a limited range of sport, appropriate children's programmes such as *Blue Peter*). And a good number of readers will have encountered the restrictions imposed on their homes by planners and architects, notably those of us brought up in council accommodation.

Against this the postmodern mentality celebrates the fact that there is no 'truth', but only versions of 'truth': this being so makes a nonsense of the search for 'truth'. In its stead the advocacy is for difference, for 'anything goes'. A consequence is that the modernist enthusiasm for genres and styles (which at one time or another would have served to situate worthwhile art and to help identify good taste) is rejected and mocked for its pretensions. From this it is but a short step to the postmodern penchant for parody, for tongue-in-cheek reactions to established styles, for a pastiche mode which delights in irony and happily mixes and matches in a 'bricolage' manner. An upshot is that postmodern architecture happily clashes received styles, famously 'Learning from Las Vegas' (Venturi, 1972; Jencks, 1984), perhaps combining Spanish-style woodwork with a Gothic façade or a ranch-style design with Venetian facings; or postmodern dress will contentedly put together an eclectic array of leggings, Doc Marten's boots, Indian necklace, waistcoat and ethnic blouse. . . .

Perhaps most noteworthy of all, postmodern culture abandons the search for 'authenticity'. Here one might list a series of cognate words which are recurrent targets of proponents of postmodern culture: the 'genuine', 'meaning', and the 'real'. Each of these terms testifies to the modernist imperative to identify the 'true'. It is, for instance, something which motivates those who seek the 'real meaning' of the music they happen to be listening to, those who look for an 'authentic' way of life which might recover the 'roots' of the 'real England' (or even of the 'real me'), those who desire to find the 'true philosophy' of the 'good life'. Against all of this, postmodernism celebrates the inauthentic, the superficial, the ephemeral, the trivial and the flagrantly artificial.

Postmodernism will have no truck with yearnings for authenticity, for two main reasons. The first is because the insistence on one 'true' meaning is demonstrably a fantasy; hence those who go looking for the 'authentic' and the 'real' are bound to fail because there can be only *versions* of the 'real'. We cannot hope to recover, say, the authentic Dickens because we read him as citizens of the late twentieth century, as, for example, people who are alert to notions such as the unconscious and child sexuality which, unavoidably, make us interpret the character of Little Nell in ways which set us apart both from the author and from his original audience.

The second asserts that the authentic condition, wherever one seeks for it, can never be found because *it does not exist* outside the imaginings of those who yearn for it. People will have it that, somewhere – round that corner, over that horizon, in that era – the real, the authentic, can be found. And, when it can be discovered, we can be satisfied at having discovered the genuine (in oneself, of one's times, of a country) which may then be set against the superficial and artificial which seem to predominate in the contemporary world of 'style', 'show' and an 'only-in-it-for-the-money' ethos. It is the contention of postmodernism that this quest for authenticity is futile.

Take, for example, the popular search for one's roots by tracing one's family back through time. Many people nowadays go to great pains to detail their family tree in order to trace their own point of origination. A common expression of this attempt to establish authenticity is the return of migrants to places from whence their forebears moved generations before. What do these seekers discover when they reach the village from which the Pilgrim Fathers fled, the Irish hamlet from which the starving escaped, the Polish ghetto from which they were driven? Certainly not authenticity: much more likely a reconstruction of the Puritan's barn-like church 'exactly like it was', a 'real' potato dinner (with cooled Guinness and fine wines if desired), a newly-erected synagogue with central heating installed and a computerised record of family histories.

You yearn to find the 'real' England? That 'green and pleasant land' of well-tended fields, bucolic cows, unchanged landscape, whitewashed cottages, walled gardens and 'genuine' neighbours that is threatened by motorway construction, housing estates and the sort of people who live in one place only for a year or so before moving on? That place where one might find one's 'real self', where one may discover one's 'roots', something of the authentic English way of life that puts us in touch with our forebears?

But look hard at English rural life – the most urbanised country in Europe – and what do we find? Agribusiness, high-tech farming, battery hens, and 'deserted villages' brought about by commuters who leave their beautifully maintained properties (which are far beyond the budgets of locals) with the central heating pre-set to come on when required and the freezer well stocked from the supermarket, to drive their Volvos to and from their town-centred offices. It is these incomers who have been at the forefront of reconstructing the 'traditional' village: by resisting industrial developments (which might have given jobs to one-time farm workers displaced by combine harvesters, tractors and horticultural science), by having the wherewithal to have the former smithy rebuilt (often as a second home – with all mod. cons), by being most active in sustaining the historical societies (which produce those wonderful sepia photographs for the village hall which show 'what life used to be like in the place we now cherish'), and, of course, by resurrecting 'traditions' like Morris dancing and village crafts such as spinning and weaving (Newby, 1985, 1987).

The point is not to mock modern-day village life but to insist that the search for an 'authentic' England is misconceived. In the late twentieth century we can only *construct* a way of life which appears to us to echo themes from another time (without the absolute hunger, poverty and hardship the majority of country dwellers had to endure). This construction of a supposedly authentic way of life is, necessarily, itself inauthentic – and ought to be recognised for what it is.

There is no authenticity; there are only (inauthentic) constructions of the authentic. Take, for instance, the tourist experience (Urry, 1990). Brochures advertise an 'unspoiled' beach, 'must see' sites, a 'distinctive' culture, 'genuine' locals and a 'taste of the *real*'. But the experience of tourism is demonstrably inauthentic, a carefully crafted artifice from beginning to end: it is the café on the beach – with well-stocked fridge full of continental beers; the customary music and traditional dancing – played on compact discs, with waiters coached in simplified steps and instructed to 'let the tourists participate'; the peasant cuisine – cooked in the microwave, stored in the freezer and combined to appeal to the clients' palates while retaining a hint of the 'local'; the obliging locals who are uncorrupted by metropolitan ways – and trained in hotel schools in Switzerland; the special tourist attractions – developed and hyped for tourist consumption. The 'tourist bubble' is intended to ensure that only pleasant experiences are undergone, that the visitors will avoid, for example, the smells and insanitary conditions endured by many of the indigenous people. Further, tourism is big business and it acts accordingly: aeroplanes must be filled, hotel rooms booked (and of a standard to meet the expectations of visitors from affluent societies, hence showers, clean bed linen, and air conditioning where appropriate), and people given a good time. All this requires arrangements, artifice, inauthenticity (Boorstin, 1962: 100–122).

Inauthenticity is also a pervasive feature of this country. Indeed, it can be argued that Britain generates an array of museum sites, architecture and amusements not merely to sustain a massive tourist industry, but also to express our 'real history'. The 'heritage industry' is centrally involved in this creation and development of our past, dedicated to rebuilding and refurbishing it in the name of evoking it 'as it really was'. Consider here examples such as the Beamish Industrial Museum in County Durham, the Jorvik Centre in York, Ironbridge and the Oxford Story. How ironic, assert the postmodernists, that so many of these tourist attractions have been arranged with a claim to make visible life 'as it once was', given that their construction unavoidably undermines claims to authenticity.

The postmodern era thus rejects all claims for the 'real': *nothing* can be 'true' and 'authentic' since everything is a fabrication. There is no 'real England', no 'real history', no 'real tradition'. Authenticity is nothing more than a (inauthentic) construction, an artifice. This being so, it follows that the recurrent and urgent question of modernists – 'what does this mean?' – is pointless. Behind every such question is an implicit idea that *true meaning* can be perceived, that, for instance, we may discover what the Bible really means, what the architect means when she designs a building in a particular manner, what it really meant to live during the Napoleonic Wars, what that girl means to suggest when she wears that sort of frock. . . .

But if we know that there is no true meaning but only different interpretations (what Roland Barthes called *polysemous* views), then, logically, we can jettison the search for meaning itself. To the postmodern temper the quest is vain, but, far from despairing at this, the suggestion is that we abandon it and instead take pleasure in the *experience of being*. For instance, you may not know how to make sense of a particular hair style, you may be bemused by each of your friends seeing it in different ways, but what the heck – enjoy the view without yearning for it to have any special meaning. The French have a word for this, *jouissance*[1], but the central idea to the postmodernist is that, where everyone knows that there is an infinity of meanings, then we may as well give up on the yearning for any meaning. As the graffiti has it, forget trying to work out what Elvis was trying to say in *Jailhouse Rock*, it's 'only rock and roll', so get up and dance and enjoy the experience.

Moreover, we intellectuals ought not to concern ourselves about this abandonment of meaning. Ordinary people themselves recognise that discovering the 'true meaning' is an unattainable dream just as clearly as we do. They too are aware of multiple meanings being generated for every situation, of the untenability of finding the authentic element. Accordingly, the people do not get uptight about finding out the real sense of the latest movie: they are quite content to enjoy it for what it was

172

to them – fun, boring, diverting, an escape from housework, a chance to woo one's partner, a night out. . . . Modernist zealots are the ones that worry about 'what it all means'; postmodern citizens gave up on that earnestness long ago, content to revel in the manifold pleasures of experience. Similarly, postmodern tourists know well enough that they are not getting an authentic experience; they are cynical about the local boutiques selling 'genuine' trinkets, about the fervent commercialism of the tourist trade, about the artificiality of an out-of-the-way location that yet manages to incorporate the latest video releases. Tourists know full well that it is all a game, but – knowing this – are still content to go on holiday and take part in the staged events, because what they want while on vacation is a 'good time', is 'pleasure', and hang any *angst* about 'what it all means' and whether or not the food, people and milieu are authentic (Featherstone, 1991: 102).

Postmodernism's emphasis on differences – in interpretation, in ways of life, in values – is in close accord with the abandonment of belief in the authentic. For instance, the postmodern outlook encourages rejection of elitisms which proclaim a need to teach children a unifying and enriching 'common culture' or the 'great tradition' of literature. All this and similar such protestations are dismissed as ideology, instances of power being exercised by particular groups over others. However, postmodern culture goes further than this: it contends that those who fear what they regard as that fragmentation of culture if people are not taught to appreciate, say, the literature and history which tells us 'what we are' and thereby what brings us as citizens together, should be ignored. On the one hand, this is because the identification of a 'common culture', whether in the Arnoldian sense of the 'best that is thought and said' or simply in the sense of 'all that is of value to our society', is usually expressive of power which can be exclusionary and impositional on many groups in our society (the 'great tradition' in English literature may not have much appeal for ethnic minorities, the working class or the young in contemporary Britain). On the other hand, however, postmodernists argue that it also presumes that people have difficulty living with fragmentation, that if things are not consistent and whole then we will experience alienation, anxiety and depression.

But the postmodern outlook positively thrives on differences and hence prospers too with a fragmentary culture. What is wrong with, for example, reading a bit of Shakespeare as well as listening to reggae music? For a long time cultural custodians have presumed to tell people what and how they ought to read, see and hear (and to feel at least a twinge of guilt when they deviated from the prescribed works and judgements). Behind this moral stewardship is a typically modernist apprehension that fragmentation is positively harmful. Against this, postmodern culture, having spurned the search for 'true meaning' ('Englishness means you are familiar with and appreciate this history, these novels, that poetry . . . '), suggests that fragmentation can be and is *enjoyed* without people getting much vexed about conflicting messages. The outcome is celebration of a plurality of sources of *pleasures* without meaning: the neon lights, French cuisine, McDonalds, Asiatic foods, Bizet, Madonna, Verdi and Gary Glitter. . . . A promiscuity of different sources of pleasure is welcomed.

Furthermore, it will be easily understood that behind the modernist apprehension about a fragmentary culture lurks the fear that the *self* itself is under threat. Such fear presupposes that there is in each of us a 'real self', the authentic 'I', which must be consistent, unified and protected from exposure to widely diverging cultural signals. How, for instance, can a true intellectual sustain her sense of self if she reads Plato and then goes dog racing? How can a real philosopher immerse himself in the giants of his discipline and then support Tottenham Hotspur Football Club? How can the integrity of the self be maintained if the same actor is exposed to role models as diverse as Clint Eastwood, Paul Gascoigne and Woody Allen?

Rather than get wrapped up trying to unravel such contradictions, postmodern culture denies the existence of an essential, true, self. The postmodern temper insists that the search for a 'real me' presupposes an underlying meaning, an authentic being, which is just not there – and hence not worthy of pursuit. Instead, the advocacy is to live with difference, in the wider society and within one's being, and to live this without anxiety about meaning, jettisoning restrictive concepts like 'integrity' and 'morality', and opting instead for pleasure.

Finally, and this is consistent with its hostility towards those who seek to reveal the 'real meaning' of things, postmodern culture lays stress on the creativity and playfulness of ordinary people. Amongst modernist thinkers there is a recurrent tendency towards offering determinist explanations of behaviour. That is, it is characteristic of modernist analyses that they present accounts of actions which privilege their own explanations rather than those of the people involved, as if they alone are capable of discerning the real motivations of those whom they study. Consider, for example, Freudian accounts which see sexuality behind so much action – whatever those studied may feel; or Marxist examinations of the world which contend that consciousness is shaped by economic relationships – whatever else subjects might say; or feminist accounts of women's experiences which frequently suggest that the analysts have privileged access to what women 'really need' – whatever the women they study may suggest.

There is from postmodernists a repeated assertion that intellectuals have no more right to recognise 'truth' than the man or woman in the street. Similarly, the widespread fear amongst intellectuals that the people are being duped, that they are being led away from the 'truth' by manipulative politicians, trashy entertainment or by the temptations of consumerism, is at once an insufferable arrogance (by what right can intellectuals claim to discern 'truth' when their own record is dubious and, from within their own ranks, intellectuals contest the 'truth' of other intellectuals?) and a nonsense given the capacities of ordinary people to see, and to create, just as effectively as any intellectual. In a world where there are only versions of truth, people have an extraordinary capacity to generate an anarchic array of meanings, and, prior even to meaning, alternative *uses* of things and experiences which they encounter (Certeau, 1984).

It will not surprise readers who have gone this far to learn that a *bête noire* of postmodernism is the claim to identify the essential features of any phenomenon. 'Essentialism' provokes the postmodernist to recite the familiar charges against arrogant modernist presumptions: that the analyst can impartially perceive the 'truth', that features hidden beneath the surface of appearances are open to the scrutiny of the privileged observer, that there is a core meaning which can be established by the more able analyst, that there are authentic elements of subjects which can be located by those who look hard and long enough.

Well, since I do not subscribe to postmodern thought, I summarily review key elements of postmodernism as an intellectual and as a social phenomenon. These include:

- the rejection of modernist thought, values and practices;
- the rejection of claims to identify 'truth' on grounds that there are only versions of 'truth';
- the rejection of the search for authenticity since everything is inauthentic;
- the rejection of quests to identify meaning because there are an infinity of meanings (which subverts the search for meaning itself);
- the celebration of differences: of interpretations, of values, and of styles;
- an emphasis on pleasure, on sensate experience prior to analysis, on *jouissance* and the sublime;
- a delight in the superficial, in appearances, in diversity, in parody, irony and pastiche;
- a recognition of the creativity and imagination of ordinary people which defies determinist explanations of behaviour.

POSTMODERNISM AND INFORMATION

But what has all this to do with information? A first, and recurrent, response comes from the postmodern insistence that we can know the world only through *language*. While Enlightenment thinkers subscribed to the idea that language was a tool to describe an objective reality apart from words, the postmodernist asserts that this is 'myth of transparency' (Vattimo, 1992: 18) because it is blind to the fact that symbols and images (i.e. information) are the only 'reality' that we have. We do not, in other words, see reality through language; rather, language is the reality that we see. As Michel Foucault put it, 'reality does not exist . . . language is all there is and what we are talking about is language, we speak within language' (quoted in Macey, 1994: 150).

An illustration of some of the consequences of this starting point at which 'language is never innocent' (Barthes, 1967: 16) can be found in literary criticism. Once upon a time critics took it as their task to discern, say, ways in which we could get a better picture of Victorian capitalism through reading *Dombey and Son*, or to examine the ethos of masculinity evidenced in the short stories of Ernest Heming-way, or to assess how D. H. Lawrence's upbringing shaped his later writing. The presupposition of critics was that one could look *through the language* of these authors to a reality behind the words (to a historical period, an ideology, a family background) and the aspiration of these critics was that they themselves would elucidate this function as unobtrusively as possible. To such intellectuals *clarity* of writing, from both artist and critics, was at a premium, since the prime task was to look through the language to a reality beyond.

Roland Barthes (1963, 1964) caused a considerable intellectual fuss in the early 1960s inside French literary circles when he attacked such assumptions in a heated debate with a leading critic, Raymond Picard. Barthes offered a reading of Racine, an icon of classical French literature, which first objected to the supposition that the meaning of Racine's words is inherently clear and, second, insisted that all critical approaches developed and drew upon *metalanguages* (Freudianism, Marxism, structuralism etc.), something which subverted any ambition of critics themselves to make clear the text (Barthes, 1966) by, for instance, making more comprehensible the historical context of its production.

The centrepiece of Barthes' objection here, of course, is that *language is not transparent*, is that all authorship, literary or otherwise, is not about looking through language to a phenomenon *out there*, but is solely a matter of *languages*, whether those of authors or critics.

The pertinence of this literary debate to our concern with postmodernism begins to become evident when we realise that Barthes and others extend their principle that language is all the reality we know to a wide variety of disciplines, from history to social science. Across a wide range they endeavour to analyse the 'phrase-regime' (Lyotard) which characterises particular subjects. As such they query the truth claims of other intellectuals and suggest alternative – postmodern – approaches to study which examine subjects as matters of language (or, to adopt the favoured word, *discourses*).

It is significant too that Barthes (1979) applied his approach to an enormous variety of phenomena in the contemporary world, from politicians, wrestlers, movies, fashion, cuisine, radio and photography to magazine articles, always discussing his subjects as types of language. Following this route taken by Barthes, we can see that, if reality is a matter of language/discourse, then everything that we experience, encounter and know is informational. Nothing is transparent or clear since everything is constructed in language and must be understood in language. In sum, one relevance of postmodernism to considerations of information is the perception that we do not live in a world about which we have information. On the contrary, *we inhabit a world which is informational*.

JEAN BAUDRILLARD

Jean Baudrillard (born 1929) elaborates principles found in Barthes and discusses them expressly in relation to developments in the informational realm. It is the view of Baudrillard that contemporary culture is one of *signs*. Nowadays just about everything is a matter of signification, something obviously connected with an explosive growth in media, but related also to changes in the conduct of everyday life, urbanisation and increased mobility. One has only to look around oneself to understand the point: everywhere we are surrounded by signs and modes of signification. We wake to radio, watch television and read newspapers, spend a good part of the day enveloped by music from stereos and cassettes, shave and style ourselves in symbolic ways, put on clothes that have sign content, decorate our homes with symbolic artifacts, add perfumes to our bodies to give off (or prevent) particular signals, travel to work in vehicles which signify (and which contain within them systems that allow the uninterrupted transmission of signs), eat meals which are laden with signification, and pass by and enter buildings that present signs to the world.

While pre-industrial societies had complex status rankings, elaborate religious ceremonies and gaudy festivals, the rigours of subsistence and the fixity of place and routine must have delimited the use of signs (cf. Baudrillard, 1983b). Nowadays we no longer mix with the same people in the same places in the same way. We interact now with strangers to whom we communicate but parts of ourselves by signs – as a passenger on a bus, or a client in a dentist's surgery, or as a customer in a bar. At the same time we receive messages from anywhere and everywhere in our newspapers, books, Walkmans or – surely most definitively – television.

It is this which is Jean Baudrillard's starting point: today life is conducted in a ceaseless circulation of signs about what is happening in the world (signs about news), about what sort of identity one wishes to project (signs about self), about one's standing (signs of status and esteem), about what purposes buildings serve (architectural signs), about aesthetic preferences (signs on walls, tables, sideboards), and so on. As John Fiske (1991) observes, that our society is sign-saturated is indicative of 'a categorical difference . . . between our age and previous ones. In one hour's television viewing, one is likely to experience more images than a member of a non-industrial society would in a lifetime' (p.58).

Baudrillard and like-minded thinkers (Virilio, 1986) go much further than just saying that there is a lot more communication going on. Indeed, their suggestion is that there are other characteristics of postmodern culture which mark it out as a decisive break with the past.

We can understand these better by reminding ourselves how a modernist might interpret the 'emporium of signs'. Thinkers such as Herbert Schiller and Jürgen Habermas acknowledge the explosive growth of signification, but they insist that, if used adroitly, it could serve to improve the conditions of existence. Such approaches perceive inadequacies in signs which, if rectified, could help to facilitate a more communal society or more democratic social relationships. What is evident in such modernist interpretations is that critics feel able to identify *distortions* in the signs which, by this fact, are in some way *inauthentic*, thereby holding back the possibility of progressing to more genuine and open conditions. For example, it is usual for such writers to bemoan the plethora of soap operas on television on the grounds that they are escapist, trivial and profoundly unreal depictions of everyday lifestyles. Tacit in such accounts is the view that there are more authentic forms of drama that may be devised for television. Similarly, modernist scholars are at pains to identify ways in which, say, news media misrepresent real events and issues – and implicit in such critiques is the idea that authentic news coverage can be achieved.

Baudrillard, however, will have neither this hankering after 'undistorted communication' nor any yearning for the 'authentic'. In his view since everything is a matter of signification then it is unavoidably a matter of artifice and inauthenticity because this, after all, is what signs are. Modernist critics will insist that there is some reality behind signs, perhaps shrouded by unreliable signs, but real nonetheless, but to Baudrillard there are only signs. As such one cannot escape inauthenticity and there is no point in pretending that one can.

For example, viewers of television news may watch with the presumption that the signs indicate a reality beyond them – 'what is going on in the world'. But on a moment's reflection we can appreciate that the news we receive is a version of events, one shaped by journalists' contacts and availability, moral values, political dispositions and access to newsmakers. Yet, if we can readily demonstrate that television news is not 'reality' but a construction of it, then how is it possible that people can suggest that beyond the signs is a 'true' situation? To Baudrillard the 'reality' begins and ends with the signs on our television screens. And any critique of these signs offers, not a more authentic version of the news, but merely *another* set of signs which presume to account for a reality beyond the signs.

Baudrillard takes this insight a very great deal further by asserting that nowadays everybody knows this to be the case, the inauthenticity of signs being an open secret in a postmodern culture. In other words, when once it might have been believed that signs were *representational* (in that they pointed to some reality beyond them), today everybody knows that signs are *simulations* and nothing more (Baudrillard, 1993a).

For example, one may imagine that advertisements might represent the qualities of particular objects in a true way. That they manifestly do not is a frequent cause of irritation to modernist critics who claim to reveal the distortions of advertisements which suggest, say, that a certain hair shampoo brings with it sexual allure or that a particular alcoholic drink induces sociability. The modernist who exposes the tricks of advertisers works on two assumptions: that he is privileged to recognise the deceptions of advertisers, to which most consumers are blind, and that an authentic form of advertising in which the advertisement genuinely represents the product is capable of being made.

Baudrillard's retort is that ordinary people are quite as knowledgeable as modernist intellectuals, but they just do not bother to make a fuss about it. Of course they realise that advertisements are, well, advertisements. They are not the 'real thing', just make-believe, just simulations. *Everybody*, and not just intellectuals, knows that Coca-Cola does not 'teach the world to sing', that Levi jeans won't transform middle-aged men into twenty-year-old hunks, or that Wrigley's chewing gum will not lead to thrilling sexual encounters. As such, we ought not to get concerned about advertising since the 'silent majorities' (Baudrillard, 1983a) are not much bothered by it.

That said, Baudrillard does assert that people do enjoy advertisements, not for any messages the advertiser might try to convey, and certainly not because they might be persuaded to go out to buy something after watching them, but simply because advertisements can bring *pleasure*. Advertising 'acts as spectacle and fascination' (ibid.: 35); just that. Who knows, who cares, what Silk Cut, Guinness or Benson and Hedges advertisements signify? We may just enjoy the experience of looking at the signs.[2]

Similarly, consider the modernist anxiety Professor Habermas manifests when he expresses concern about the packaging of politics in contemporary democracies. To critics such as Habermas the manipulation of political information is deplorable, the public being misinformed by the politicians' and their PR advisers' knavery (rehearsals, briefings, staged events, off-the-record discussions, make-up and clothing chosen to project a desirable image, media consultants playing a disproportionate role in presenting policies and their ministers). The appeal of the critics here, explicit or not, is that politicians ought to be honest and open, truthful and direct, instead of hiding behind misleading and mendacious media 'images'.

Baudrillard's response to this modernist complaint would take two forms. On the one hand he would insist that the dream of signs which represent politics and politicians in an accurate way is but a fantasy. Unavoidably the media will only be able to show certain issues, particular personalities, and a limited range of political parties. If for no other reason, the limitations of time mean that political coverage is restricted to certain issues and political positions. Add to that the disposition of politicians to press to have the most favourable arguments for their own positions presented, then it is easy to understand that the difficulties of exactly representing politics through media are insuperable. In Baudrillard's view the fact that the media must put together a presentation of politics for the public means that any alternative presentation can be nothing but just another simulation. On the other hand, Baudrillard would assert that, since everyone knows this, then no one is much bothered since the signs are ignored. We all know that they are artificial, so we just enjoy the spectacle and ignore the messages, knowingly reasoning that 'it's just those politicians on the television again'.

Logically this knowledgeability of the public heralds what one might describe as the death of meaning. If people realise that signs are but simulations, and that all that can be conceived are alternative simulations, then it follows that anything – and nothing – goes. Thus we arrive at Baudrillard's (1993b) conclusion that 'we manufacture a profusion of images *in which there is nothing to see*. Most present-day images . . . are literally images in which there is nothing to see' (p.17). If the 'masses' recognise that signs are just simulations, then we are left with a profusion of signs which just do not signify. We have signs without meaning, signs which are 'spectacular' (Baudrillard, 1983a: 42), to be looked at, experienced and perhaps enjoyed, but without significance. This, indeed, is a postmodern world.

These examples come from media, the obvious domain of signification and an area that most readily springs to mind when one thinks of an information explosion. However, it is important to realise that Baudrillard contends that the society of spectacle and simulation reaches everywhere, and much deeper even than an enormously expanded media.

To better appreciate this, recall that everything nowadays is a sign: clothing, body shape, pub décor, architecture, shop displays, motor cars, hobbies – all are heavily informational. Again, modernist writers tend to examine these things in terms of an underlying or potential authenticity, for example assuming that there is a natural body weight for people of a given size and build, or that shop displays can be set out in such a way that customers can find what they want in a maximally convenient and unobtrusive way. However, Baudrillard rejects these approaches on the grounds that the modernist search for the authentic is misconceived since all these signs are *simulations rather than representations*.

For instance, body shape now is largely a matter of choices and people can design the signs of their bodies. If one considers the plasticity of body shape today (through diets, exercise, clothing, or even through surgery) then one gets an idea of the malleability of the human body. Now the modernist would respond to this in one of two ways: either the obsession with body shape is condemned as leading people away from their 'true' shapes (and bringing with it much anxiety, especially for young women) or people are seen as having an inappropriate body shape to sustain their 'true' health (and ought perhaps to eat less). Either way the modernist appeal is to an authentic body shape beyond the distortions induced by inappropriate role models or over-indulgers who ignore expert advice on the relations between diet and health.

But there is no authentic body shape, not least because, in the late twentieth century, we are all on a permanent diet (in that we all selectively choose from a cornucopia of foods); because experts disagree amongst themselves about the linkages between health and body shape; and because, in an era of choices, there is a wide variety of body shapes to be chosen. In these circumstances there is just a range of inauthentic body shapes, just simulations which represent neither the 'true'/ideal body shape nor a deviation from it. They just *are*, signs without significance. The test of this thesis is to ask, what does body shape signify nowadays? What, for instance, does a slim body signify in 1995? Beauty? Anorexia? Narcissism? Health? Obsession? Body shape has lost its power to signify. Having done so it is a sign to be experienced rather than interpreted.

Baudrillard is echoing here a strong social constructivist view of signs. That is, if phenomena are socially created then they are simulations with no 'reality' beyond themselves. This accounts for Baudrillard's famous claim that Disneyland does not represent, symbolically, the real America which is outside the entertainment centre (a typically modernist argument, that Disney is a mythological representation of American values, whereby visitors are surreptitiously exposed to ideology while they're busy having fun). On the contrary, Disney is a means of acknowledging the simulation which is the entirety of modern America: *everything* about America is artifice, from small town main streets to city centre corporate offices. This, proclaims Baudrillard, is all the *hyper-real*, where signs refer to nothing but themselves. As he arrestingly remarks: 'Disneyland is presented as imaginary in order to make us believe that the rest is real, when in fact all of Los Angeles and the America surrounding it are no longer real, but of the order of the hyperreal and of simulation' (1983b: 25).

In the postmodern era the distinction between the real and the unreal, the authentic and the inauthentic, the true and the false, has collapsed: when all is artifice such certainties have to go. Thus the 'historic' town, the 'seaside resort' and the 'fun' city, are hyper-real in that they have no relationship with an underlying reality. They are fabrications with no authenticity outside of their own simulations. So it is fatuous to go, with the modernist, in search of the 'real' that is imagined to be found in Blackpool Tower because there is no authenticity behind these signs. These inauthentic monuments are all that there is. They are the hyper-real, 'the generation by models of a real without origin or reality' (Baudrillard, 1988b: 166).[3]

It follows that, where 'the real is abolished' (Baudrillard, 1983a: 99), the meaning of signs is lost (it is 'imploded'). Nonetheless, we ought not to worry about this, because we always have to recall the postmodern dictum that audiences are subversive of messages anyway. A while ago modernists got themselves into a lather about 'couch potato' television viewers and tourists who visited historical sites, took a photograph, and then, having 'done it', were gone without appreciating the 'real thing'. But how much this underestimates the creativities of ordinary folk – the TV viewer is in fact constantly active, switching channels with enthusiasm, chatting to her pals, using the telephone or shouting out irreverent and irrelevant comments, and the tourist is doing all sorts of things when walking round the Natural History Museum, day-dreaming, wondering why the guide reminds him of his brother, planning dinner, chatting up the girls, musing whether diplodocus ever got toothache. . . . Given such resistance, as it were, to the intended signs, we can conclude that postmodern audiences are a far cry from the 'cultural dopes' modernists so feared, so far indeed that they see and hear *nothing*, just experience the spectacles which characterise the contemporary.

GIANNI VATTIMO

Italian philosopher Gianni Vattimo (1989) contends that the growth of media have been especially important in heralding postmodernism. In brief, the explosive growth of information from here, there and everywhere has undermined modernist confidence in 'truth' and 'reality'. Vattimo suggests that, while on the intellectual front Enlightenment tenets have been successfully challenged by, for example, alternative historical interpretations, so too has the spread of media undermined any more general commitment to a single way of seeing.

It used to be common for modernist thinkers, of Left or Right, to bemoan the development of 'mass society' where people would become herd-like, indoctrinated by media which put out a diet of homogeneous entertainment and propaganda.

Readers familiar with the writing of Frankfurt School Marxists will recognise this pessimistic vision, but conservative critics such as T. S. Eliot and Frank and Queenie Leavis felt much the same about the likely effects of film, radio and mass circulation newspapers (Swingewood, 1977).

Against this, Vattimo argues that the proliferation of media has given voice to diverse groups, regions and nations, so much so that audiences cannot but encounter many 'realities' and 'perspectives' on issues and events. Nowadays 'minorities of every kind take to the microphones' (Vattimo, 1989: 5), thereby disseminating world views which lead to a collapse in notions of the 'true'. From this comes freedom because, says Vattimo, the belief in reality and its associated persuasive force ('you must do this because it is true') is lost.

Differences come to the forefront of everyone's attention as multiple realities (sexual, religious, cultural, aesthetic . . .) get time on the airwaves. Bombarded by the very diversity of signs, one is left confused and shaken, with nothing sure any longer. The result, however, is actually liberatory and definitively postmodern, with experience taking on the 'characteristics of oscillation, disorientation and play' (Vattimo, 1989: 59).

Here Vattimo finishes up in pretty much the same position as Baudrillard. A multiplicity of signs paradoxically subverts the sign's capacity to signify and people are left with spectacle, non-meaning and freedom *from* truth.

MARK POSTER

Mark Poster forwards the proposition that the postmodern age is distinguished from previous societies because of what he designates a 'mode of information'. This suggestion of fundamental change emanating from developments in information is especially interesting both because of its elaboration of themes found in Baudrillard and because of its emphasis on the novelty of the postmodern era.

Poster's claim is that the spread of information technologies, and hence electronic mediated information, has profound consequences for our way of life and, indeed, for the ways in which we think about ourselves, because it alters our 'network of social relations' (Poster, 1990: 8). Elaborating this principle, he proposes a model of change based on different types of 'symbolic exchange' (p.6) which has three constituents:

- the era of *oralism* when interaction was face-to-face. Then the way of life was fixed and unchanging, the self embedded in the group, and signs *corresponded* to this settled way of life, with symbolic exchange a matter of articulating what was already known and accepted by the community.
- the era of *written exchange*, when signs had a *representational* role and in which the self was conceived to be rational and individually responsible.
- the era of *electronic mediation*, when signs are matters of informational *simulations*, with their *non-representational* character being critical. Here the self is 'decentred, dispersed, and multiplied in continuous instability' (p.6), swirling

in a 'continuous process of multiple identity formation' (Poster, 1994: 174) since the 'flow of signifiers' is the defining feature of the times rather than signs which indicate a given object.

What Poster suggests is that once people said and thought what was expected of them, later they developed a strong sense of autonomy and used writing especially to describe what was happening outside of themselves in the world, and then, in the postmodern present, the spread of simulation has shattered previous certainties. No longer able to believe in a 'reality' beyond signs, the self is left fragmented, unfocused and incapable of discerning an objective reality. Yet despite the dislocation this brings about, Poster sees it as emancipatory because the 'crisis of representation' (1990: 14) results in a plethora of signs which do not signify, something which at last frees people from the tyranny of 'truth'.

JEAN-FRANÇOIS LYOTARD

It is especially appropriate at this point to consider the work of Jean-François Lyotard since his work has been particularly concerned to demonstrate how truth claims have been subverted by postmodern developments. Moreover, Lyotard goes about his task by centring attention on informational trends, arguing that it is changes here which give rise to the scepticism towards truth claims which characterises postmodern culture. In addition, Lyotard provides a revealing contrast to the previous three thinkers since he arrives at remarkably similar conclusions while approaching from a different starting point. That is, while Baudrillard, Vattimo and Poster give emphasis to the rapid growth in signs (especially in media), Lyotard starts his analysis with a concern for changes in the role and functions of information and knowledge at a more general and simultaneously deeper level.

This French philosopher argues that knowledge and information are being profoundly changed in two connected ways. First, increasingly they are produced only where they can be justified on grounds of efficiency and effectiveness or, to adopt Lyotard's terminology, where a *'principle of performativity'* prevails. This means that information is gathered together, analysed and generated only when it can be justified in terms of utility criteria. This may be conceived of as a 'systems' orientation which determines what is to be known, the 'programme' of the 'system' insisting that information/knowledge will only be produced when it is of practical use.

Second, Lyotard argues that knowledge/information is being more and more treated as a *commodity*. Endorsing a theme we have already seen to be prominent in the work of Herbert Schiller, he contends that information is increasingly a phenomenon which is tradable, subject to the mechanisms of the market which has a determining effect on judging performativity.

The consequences of these twin forces, of which there are several, are enormous, sufficient even to announce the emergence of a postmodern condition. First, the principle of performativity when applied means that information/knowledge that cannot be justified in terms of efficiency and effectiveness will be downgraded or even abandoned. For example, aesthetics and philosophy cannot easily be justified in terms of performance while finance and management are straightforwardly defended. Inexorably the former suffer demotion and the latter promotion, while within disciplines research in areas that are defensible in terms of use will be treated more favourably than others. For instance, social science investigations of technology transfer have practical implications for markets and hence are seen as worthy of support from research funding bodies such as the ESRC (Economic and Social Research Council) the mission (*sic*) of which now demands that the research it sponsors must contribute to the competitiveness of industry. Conversely, the social scientist whose interest is in the exotic or impractical (as judged by performativity criteria) will be sidelined. As a government minister, Norman Tebbit, put it in the early 1980s when called upon to justify switching funds from arts, humanities and social sciences to the more practical disciplines, money was to be taken away 'from the people who write about ancient Egyptian scripts and the pre-nuptial habits of the Upper Volta valley' and given to subjects that industry thought useful.

Second, knowledge development is shifting out of the universities where, traditionally, a cloistered elite had been ensconced with a vocation to seek the 'truth'. Challenging the dominance of the traditional university is an array of think tanks, research and development sections of private corporations and pressure groups which generate and use information/knowledge for reasons of efficiency and effectiveness. For instance, commentators now speak of the 'corporate classroom' which is as large and significant as universities and colleges inside the United States. It is easy to list a roll-call of some of the major players: Bell laboratories, IBM's R&D sections, and ICI's employment of hundreds of PhDs appear to many observers to be 'just like a university' – except that they have different priorities and principles which guide their work.

Moreover, that personnel move with increasing ease between universities and these alternative knowledge/information centres indicates that higher education is being changed from within to bring it into line with performativity measures. Any review of developments in higher education in any advanced economy highlights the same trends: the inexorable advance of the practical disciplines and the retreat of those that find it hard to produce 'performance indicators' which celebrate their utility. The biggest boom subject in British higher education over the last decade, by far, has been Business and Management; every British university now boasts a clutch of sponsored professorships – in a restricted range of disciplines; it is becoming increasingly common for universities to offer special training programmes for corporations and even to validate privately created courses; there are sustained pressures to make education 'more relevant' to the 'real world' of employment by inducting students in 'competencies' and 'transferable skills' which will make them more efficient and effective employees.

Lyotard extends this argument to the whole of education, insisting that it is motivated now by criteria such as 'how will it increase my earning potential?' and 'how will this contribute to economic competitiveness?' This is a transformation that impacts not just on schools and universities, but also changes the very conception of education itself. In the view of Lyotard, performativity criteria mean there will be a shift away from education perceived as a distinct period in one's life during which one is exposed to a given body of knowledge towards on-going education throughout one's life, to be undertaken as career and work demands so dictate.

Third, and a consequence of this redefinition of education, established conceptions of truth are undermined, performativity and commodification leading to definitions of truth in terms of utility. Truth is no longer an unarguable fact and the aspiration of the university; rather truths are defined by the practical demands placed on the institution. This development is a defining element of postmodernism, since the replacement of TRUTH with a 'plurality of truths' means that there are no longer any legitimate arbiters of truth itself. The upshot is that truth is merely a matter of a 'phrase regime', something defined by the terms in which one talks about it.

In this respect the undermining of traditional universities (which had been regarded as definers of legitimate knowledge) and, connectedly, intellectuals, is central (Bauman, 1987). According to Lyotard intellectuals must pursue knowledge in terms of a 'universal' ambition, be it humanity, the people, Enlightenment, the proletariat or whatever. It scarcely needs saying that many intellectuals resist the rise to prominence of performance-defined expertise, scorning those guided in the development of information/knowledge by practicality as 'mere technicians'. Against these who function only within the boundaries of an 'input/output . . . ratio relative to an operation' (Lyotard, 1993: 4) intellectuals usually aspire to research, write and teach for a much wider constituency.

However, as we near the century's end, the intellectuals' justifications sound increasingly hollow within and without education. This is partly a result of lack of resources, the distribution of which makes division difficult and the inevitable squabbling demeaning. More fundamentally, however, it is a consequence of the collapse of intellectuals' *raison d'être* since at least the post-Second World War period. The point is that it is precisely the intellectuals' claims to have privileged access to truth, to have a totalising vision, which have been destroyed. Lyotard, the one-time Communist, identifies the collapse of Marxism in the wake of revelations about the Gulag amidst its manifest economic inadequacies, as especially significant in this regard. Marxism's claim for universal truth no longer holds any credibility, and neither do the superiorities of other intellectuals, whether they be couched in terms of the value of the Classics, History or Great Literature. Today, if one argues that a particular discipline, vocation or aspiration is superior to others, then it is widely regarded as a partisan proposition, a 'phrase regime' with no more legitimacy than anything else. As degrees in Tourism, Public Relations and Business Administration proliferate in British universities, any proposal from other academics that their disciplines have more value because they offer students greater access to truth, more understanding of the 'human condition', or more profundity is greeted with at least derision or, more commonly, the accusation that this is expressive of an unworldly and useless snobbery.

The solid grounds on which intellectuals once belittled 'technicians' have turned to sand – and this is widely appreciated. No one, attests Lyotard, has recourse to the Enlightenment justification for education any more – that more education leads to better citizens – though this was once a popular universalist claim. History has destroyed its legitimacy: nowadays 'no one expects teaching . . . to train more enlightened citizens,' says Lyotard (1993), 'only professionals who perform better . . . the acquisition of knowledge is a professional qualification that promises a better salary' (p.6).

Fourth, and finally, performativity criteria applied to information/knowledge change ideas about what is considered to be an educated person. For a long while to be educated meant to be in possession of a certain body of knowledge; with computerisation, however, it is more a matter of knowing how to access appropriate data banks than of holding the information in one's head. In the postmodern age performativity decrees that 'how to use terminals' is more important than personal knowledge. Therefore, competencies such as 'keyboard skills' and 'information retrieval' will displace traditional conceptions of knowledge (and student profiles will certify that these and other competencies have at least equivalent recognition to more orthodox academic attainments) as 'data banks [become] the Encyclopaedia of tomorrow' (Lyotard, 1993: 51).

Moreover, data banks and the competencies to use them further undermine the truth claims of traditional elites. They announce 'the knell of the age of the Professor', since 'a professor is no more competent than memory banks in transmitting established knowledge' (ibid.: 53) and, indeed, is poorer at using that in a versatile and applied manner than the *teams* of employees that are increasingly required in the world of work (in preparation for which students will be trained and credited in 'skills' such as 'working in groups', 'leadership' and 'problem-solving').

What all of this amounts to, of course, is the relativism of knowledge/information. To Lyotard, performativity, commodification and the manifest failure of 'grand narratives' have resulted in a refusal of all notions of privileged access to truth. Some intellectuals might despair at this, but Lyotard (1993) considers that it can be liberatory because the decline 'of the universal idea can be free thought and life free from totalizing obsessions' (p.7). With this, again, we are deep within postmodern culture.

CRITICAL COMMENT

Postmodern thinkers have interesting and insightful things to say about the character and consequences of informational developments. I do not think anyone can try seriously to understand the contemporary world without some awareness of the centrality and features of signification today (Baudrillard), without some consideration of changes in modes of communication (Poster), without some recognition of the diversity and range of world views made available by modern media (Vattimo), and without some attention to the import of performativity criteria and commodification for the informational realm (Lyotard).

However, postmodern thought's determination to relativise knowledge, to insist that there is no truth but only (an infinity of) versions of truth, has to be jettisoned. Not least because it is inherently contradictory, manifesting the ancient Cretan paradox that 'all men are liars'. How can we believe postmodernism's claims if it says that all claims are untrustworthy? This is, to use the blunt term of Ernest Gellner (1992), 'metatwaddle' (p.41), something which fails to acknowledge that there is truth beyond the 'discourses' of analysts. That is, against postmodern thinkers one may pose a reality principle, that there is a real empirical world beyond one's imaginings (Norris, 1990). This is not to say that there is TRUTH out *there*. Of course it must be established in language, but this does not subvert the fact that truth is more than just a language game. Moreover, though we may never grasp it in any absolute and final sense, we can develop more adequate versions of reality by demonstrating better forms of argumentation, more trustworthy evidence, more rigorous application of scholarship, and more reliable methodological approaches to our subjects. If this were not so, then the revealed 'truth' of the religious zealot must be put on a par with that of the dispassionate scholar (Gellner, 1992), a collapse into relativism with potentially catastrophic consequences.

It is this insistence on absolute relativism that reduces Baudrillard's commentary often to downright silliness. To be sure, he is right to draw attention to the manufacture of news and to remind us that this construction of signs is the only reality say, of events in Bosnia that most of us encounter. However, it is when Baudrillard continues to argue that news is a simulation *and nothing more* that he exaggerates so absurdly as to be perverse. He is absurd because it is demonstrably the case that all news worthy of the term retains a representational character, even if this is an imperfect representation of what is going on in the world, and this is evidenced by comparing alternative news presentations of the same issues and events and also by realising that there is indeed an empirical reality towards which newsgatherers respond. It is surely necessary to retain the principle that news reports are, or can be, representational so that one can, with reliability if scepticism, judge one news story as more accurate, as more truthful, than another, for example the coverage of the BBC news teams compared to that offered by the Serbian aggressors. As we undertake this comparative task, we also realise that we are engaged in discriminating between more and less adequate – more or less truthful – representations of events, something which gives the lie to the postmodern assertion that there is either a 'truth' or an infinity of 'truths'.

More urgent than retaining the principle that news coverage has a representational quality, however, is the need to remind ourselves that the news reports on an empirical reality. It may not do this terribly well, but unless we remember that there is a real world we can finish in the position of Baudrillard (1991) when he insisted, before the shooting started, that the Gulf War never happened since it was all a media simulation or, after the event, merely a war-game simulation of nuclear war (Baudrillard, 1992: 93–94).

This is by no means to deny that the Gulf War was experienced by most of the world solely as an informational event, nor that this was the most extensively reported war of all time, nor that most media coverage was deeply partisan and even propagandistic (Mowlana *et al.*, 1992). On the contrary, it was just because the news of the Gulf War was widely perceived to be flawed that we may point to the possibility of representational news being produced about it, and to the possibility of discriminating between types of coverage to identify the more reliable from the less. This is the value, for instance, of journalists such as Hugo Young (1991) who, during the battles, warned his readers to beware 'the illusion of truth' that came from 'wall-to-wall television' reportage. Alerting his readership to the fact that 'nobody should suppose that what they hear in any medium is reliably true', he continues to identify the crucial issue: 'that we are consigned to operate with half-truths' and demands that 'we journalists should hang on to it'. That is, we ought to be sceptical indeed of the reportage, but this must make us all the more determined to maximise access to reliable information (cf. Pilger, 1991a). Part of that was the information – ignored by Baudrillard in his enthusiasm for 'simulations', but made available in media at the time – that 'as many as 100,000 Iraqis .may have been killed or wounded . . . in a concentration of killing . . . unequalled since Hiroshima' (Flint, 1992).

Further, Baudrillard's strictures on the implausibility of seeking the authentic have an easy appeal in an age of 'virtual reality' technologies which can precisely simulate experiences such as flying an aircraft and in a society such as England where the heritage industry is determinedly reconstructing historical landscapes. But, once again, the problem with Baudrillard is his rampant relativism which refuses to discriminate between degrees of authenticity. To suggest that this may be undertaken is not to say there is some core, some eternally genuine article, but it is to argue that one can, through critique, discriminate between phenomena to identify the more authentic from the less so.

Finally, Baudrillard's assertion that we are left only with 'spectacles' which are to be experienced but not interpreted reflects again his disdain for empirical evidence. It is undeniable that, in the contemporary world, we are subject to a dazzling array of fast-changing signs, but there is no serious evidence that this results in the abandonment of meaning. It does make clear-cut interpretation of signs exceedingly difficult, but complexity is no grounds for asserting that, with interpretation being variable, interpretation itself is lost. People are not yet sign-struck, not yet the gawping 'silent majorities' Baudrillard imagines (Kellner, 1989a).

Mark Poster echoes a good deal of Baudrillard's assertions, and much the same objections to his work are pertinent. In addition, however, one can remark on features of his historical analysis. Poster's tripartite history – oralism, writing, electronic exchange – is deeply technological determinist and subject to the familiar objection that it is historically cavalier (Calhoun, 1993).

188

Gianni Vattimo is, of course, correct to draw attention to the multi-perspectivism that the expansion of media can bring. Television has brought to our homes experiences from other cultures and, indeed, from within our own society (Meyrowitz, 1985), which can challenge and disconcert. However, a glance at the mountain of empirical evidence must reveal the marked limitations of this perspectivism since it shows clearly that some perspectives are a great deal more exposed than others (Tunstall, 1977).

To say that Hollywood dominates the world's movies, that American television accounts for large chunks of most other nations' programming, or that rock music originates in the main in London, Los Angeles and New York, is not to argue that alternative perspectives are ignored. Quite the contrary, it is easily conceded that other cultures are noticed and even given voice here – consider, for instance, rap music or the urban movies which might show life through the eyes of ethnic minorities.

But to accept that media have opened out to include other ways of seeing, at the same time as they have expanded exponentially, is by no means the same as agreeing that they offer 'multiple realities'. On the contrary, it is surely the case, as scholars such as Herbert Schiller demonstrate time and again, that what perspectives are to be included are subject to ideological and economic limits. That is, while some cultures may be given voice, it is an inflected one which is, as a rule, packaged in an appropriate and acceptable way for media corporations and, above all, it must be – or be made – marketable, something which limits the potential of, say, Chinese or Ukrainian ways of seeing to get much air time.

A fundamental objection to Vattimo, as well as to other postmodern commentators, is that his account is devoid of an empirical analysis which endeavours to assess the realities of media output. His point that a profusion of media has led to inclusion of some 'alternative realities' is well made and obvious. However, analysis needs to go beyond this truism, to demonstrate the variation in perspectives (and the discernible limits placed on that which gets access to media) as well as the differential exposure of these perspectives.

This absence is also noticeable in the work of Lyotard, though his account of the influence of performativity criteria and the commodification of information/knowledge is revealing. One can readily discern, in an enormous range of spheres, the influence of performativity and commodification: in publishing, where 'how to' and 'blockbusters' predominate; in television, where the 'ratings' are the critical measure of success since these bring in advertising revenue; in research and development activity where 'marketable solutions' are sought by investors, where scientists are compelled to sign copyright waivers, and where 'intellectual property' is protected in patent submissions . . . .Above all, perhaps, Lyotard refocuses attention on the educational sphere, surely a quintessential, but often downplayed, element of the 'information society', to demonstrate the intrusion of performativity criteria and the increased commercialisation of affairs (Robins and Webster, 1989).

The main problem with Lyotard, however, is that he concludes from all of this that the reliability of all knowledge is lost and that an appropriate response is to

189

celebrate our release from the 'tyranny' of truth. This gay abandon appears oblivious to the power and interests which have guided and continue to direct the spread of performativity and commodification. Moreover, were one to identify the processes and agencies of power and interest this would be to describe a reality which implies the possibility at least of alternative ways of arranging matters: 'This is as it is and why it is so – we can make it different.' In short, it would be to uphold the Enlightenment ideal of pursuing an alternative, and better, way of life.

A POSTMODERN CONDITION

My objections to postmodern thought do not mean that I deny the reality of something one might call a postmodern condition. I do not believe that there is any hard evidence for a collapse of meaning, still less a widespread denial of reality (a dip into the annual *British Social Attitudes* surveys cast doubts on the postmodern enthusiasm for discovering *difference* everywhere), but one can concede that signs of what may be taken to be postmodern lifestyles are manifested in hedonistic, self-centred (and maybe even de-centred) behaviours, scepticism about definitive 'truth' claims, ridicule and hostility towards 'experts', delight in the new, pleasure in experiences, and a penchant for irony, pastiche and superficiality.

It is the way in which this is explained that sets apart major commentators on the postmodern condition, such as David Harvey, from postmodern scholars. Essentially these modernist thinkers argue that the postmodern condition is a product of long-term developments in capitalist relations. That is, there are under-lying features that may be identified by diligent scholars which help account for the changes we have come to call postmodernism.

Some such thinkers hesitate to suggest a definite historical cause of the post-modern condition. For instance, Fredric Jameson (1991) refers only to postmodernism as the 'cultural logic of late capitalism'. To Jameson realist culture was the correlate of market capitalism, modernist culture (as in Surrealism) in accord with monopoly capitalism, and now postmodernism is the culture that has most affinity with consumer capitalism. Scott Lash and John Urry (1987) present a similar mode of analysis, arguing that an emergent 'service class' of educated, career-oriented, individualistic and mobile people with little sympathy for ties of 'community' and 'tradition' has an 'elective affinity' with postmodern lifestyles.

David Harvey (1989b) does not hesitate to identify a stronger causal connection. In his view the features of postmodernism are the result of changes in capitalist accumulation. Bluntly, the flexibility which we associate with contemporary capi-talism – the adaptability of employees, the capacity of companies to innovate, the acceleration of change itself – gives rise to postmodern culture. To Harvey the post-war Fordist era offered standardised products manufactured in standardised ways; today post-Fordism prevails, offering choice, variety and difference from an economic system beset by crisis, facing new circumstances (microelectronic tech-nologies, world-wide competition, globalisation), and eager to find solutions in 'flexible production' and its essential correlate 'flexible consumption'. Postmodern culture is the outcome of these trends; as Harvey (1989b) writes:

> the relatively stable aesthetic of Fordist modernism has given way to all the ferment, instability, and fleeting qualities of a postmodernist aesthetic that celebrates difference, ephemerality, spectacle, fashion, and the commodification of cultural forms.
>
> (p.156)

Postmodernism accords, in other words, with the transition from Fordism to post-Fordism which we discussed in the previous chapter.

Interestingly, Daniel Bell, coming from a quite different starting point to that of David Harvey, shares a willingness to explain the postmodern condition as, in part at least, a consequence of 'the workings of the capitalist economic system itself' (Bell, 1976b: 37). Bell suggests that the very success of capitalism in generating and sustaining mass consumption, giving people cars, fashions, TVs, and all the rest, has led to a culture – he did not yet call it postmodern in the mid-1970s, but that is what it amounted to – of pleasure, hedonism, instant gratification and the promotion of experience over meaning (cf. Bell, 1990) which, paradoxically, is at odds with the sobriety and efficiency-directed value system that contributed to the startling success of capitalism in the first place.

These accounts of the postmodern condition offer rigorous historical analyses and bring forward a wealth of empirical information to provide substance to their arguments. But, of course, a determined postmodernist thinker can dismiss them all as pretentious 'grand narratives', with Harvey interpreting the postmodern condition as the working out of the inner logic of capitalist forces and with Bell coming from a committed modernist position which regards the postmodern as a culture decidedly inferior to what went before.

To the postmodernist these accounts are unacceptable because they presume to see the truth where there is no truth to be found. Harvey, for instance, claims to see beneath the surface of postmodern culture to an underlying, but determining, economic reality, presenting a vision that is said to emanate from his own commitment to Marxist principles and which relegates those he studies – the postmodern subjects – to 'cultural dopes' because they fail to see the hidden forces of capitalism with the learned professor's clarity. To the postmodernist Harvey's is but one reading, one interpretation amongst an infinity of possibilities, and one which is rather noxious at that (Morris, 1992).

None of these studies is beyond criticism, not least by those who can indicate shortcomings, absences and even prejudices in the authors. For example, David Harvey himself would concede that his book might have benefited from a more sensitive appreciation of feminism (Massey, 1991). However, from admission of the value of critique to endorsement of the postmodern dogma that everything is but an interpretation is an unacceptable leap because in between is the matter of substantive analysis. We can readily agree that each account is partial, but it cannot be dismissed – or seen as no more than equal to any other 'reading' – on that account, because one must *demonstrate* how some accounts are more, and others less, partial.

CONCLUSION

There are indications that postmodernism has lost some of the appeal it enjoyed towards the end of the 1980s. This does not seem surprising given the flaws inherent in the postmodern project and some of the more preposterous declarations made by devotees such as Jean Baudrillard. Nonetheless, to say this is not to deny the heuristic value of a good deal of postmodern comment, not least that it helps us consider more seriously consequences of living through a period of explosive informational growth. The emphasis on the sign and signification, on simulation and inauthenticity, on the transformative power of performativity criteria applied to information and knowledge, and acknowledgement of the import of electronically mediated information, are all useful to students of the 'information revolution'.

However, it is extremely doubtful that 'we are entering a genuinely new historical configuration' (Crook *et al.*, 1992: 1). Quite the contrary, most of the postmodern condition's characteristics are explicable in terms of ongoing, if accelerating, trends, ones identified and explained by modernist thinkers such as Herbert Schiller, Jürgen Habermas, Anthony Giddens and David Harvey. Like post-industrial theory, postmodernism proclaims a new primacy of information and with it the arrival of a fundamentally different sort of society. And also as with post-industrialism, the proclamation cannot be sustained in face of sustained empirical and theoretical scrutiny.

9

INFORMATION AND URBAN CHANGE: MANUEL CASTELLS

Manuel Castells (born 1942) is Professor of Planning at the University of California, Berkeley, and one of the most eminent of contemporary writers on urban change. In *The Informational City* (1989), his concern is with subjects we have discussed in earlier chapters of this book, namely post-Fordist tendencies (Chapter 7) and the emergence of postmodern culture (Chapter 8). Consideration of Castells' analysis of changes in information and the urban environment is an especially useful way in which to bring aspects of these together. I want to focus on several dimensions of what he calls the *informational city* to examine two issues in particular: first, changes in class structures of cities which stem from restructuring processes undertaken to meet the challenges of a globalising economy, and, second, associated cultural developments which some have suggested announce the arrival of the postmodern city. This focus does mean that I shall ignore a good deal of *The Informational City*'s wider arguments that concern the relations between technological innovation, socio-economic realignment and changes in locations and places. Moreover, I shall feel free to extend beyond his own comments on stratification in informational cities and their cultures to include consideration of other contributors and even some rather speculative comment of my own.

THEORY AND THESIS

It helps an understanding of Manual Castells to know that at the time he wrote his first major book, in the early 1970s, he was a Marxist scholar. Indeed, *The Urban Question* (1972) carried the subtitle, *A Marxist Approach*. We may appreciate still more of Castells when we note that he wrote this book while teaching and researching in the University of Paris where he was heavily influenced by a particular version of Marxist thought, that of the structuralist type professed by Louis Althusser.

By the time of *The Informational City* in the late 1980s Castells, like many of his contemporaries, had followed a path to what might be called 'post-Marxism', by which I mean that much of the radical political fire had gone from his work (cf. Castells and Hall, 1994). Despite this a good deal of the conceptual apparatus of Althusserian Marxism remained.

Marxian legacies *per se* do not lessen the value of Castells' analysis. I made clear in Chapters 5 and 6 that an appreciation of the significance of information in the modern world can scarcely do without such Marxian concepts as commodification and corporate capitalism. However, I shall argue that Manuel Castells' indebtedness to the particular type of analysis developed by Althusser, notably when it comes to his account of technological innovation, manifests limitations in his approach.

Castells' main concern is with changes since the end of the Second World War, and especially since the 1970s, in the United States and beyond. His central thesis is that a combination of capitalist restructuring and technological innovation is the major factor transforming society and hence urban and regional terrains.

To highlight economic reorganisation and technological development as key factors inducing change is a familiar enough proposal these days, so much so that some might wonder what makes Castells noteworthy. However, it is worth taking time to consider Castells' ideas in some depth, for two opposing reasons. The first is that it lets us understand better the theoretical premises underpinning a form of explanation of change which is indeed commonplace at the moment. To this degree Castells' approach is notable precisely because of its orthodoxy. The second reason, which I explore later in this chapter, is that Castells' thoughts on the consequences of change for relations in space are arresting and provocative.

Let us look first at the theoretical approach. Castells distinguishes the *capitalist mode of production* from the *informational mode of development*. The former is a recognisably Marxist term, unambiguously identifying a system of production organised on market principles. That is, the capitalist mode of production is premissed on the search for profit, private ownership of property (though capital may be held by the state), competition between participants, marketability as a key determinant of what gets made and remains available, and growth (capital accumulation) as a major goal of capitalist enterprise.

Distinguished from the capitalist mode of production, which is regarded by Castells as a way of organising a social system, is the *mode of development*, which is presented as a means of generating a given level of production. According to Castells, different societies operate with different modes of development and nowadays it is information processing that announces 'the rise of a new technological paradigm [which] herald[s] a new mode of development' (Castells, 1989: 12). This *informational mode of development* is a new 'socio-technical paradigm', the main feature of which is 'the emergence of information processing as the core, fundamental activity conditioning the effectiveness and productivity of all processes of production, distribution, consumption, and management' (p.17).

We might pause on this distinction for a while to reflect on its implications. The distinction between ways of organising societies (the mode of production) and means of achieving productivity (the mode of development) is, in more direct terminology, the difference between social arrangements and technical requisites. What Manuel Castells is telling us here is that, while social relations and technical developments are closely connected, they are also independent factors in determin-

194

ing change. Indeed, he is explicit in saying that 'modes of development evolve according to their own logic' (p.11), hence the informational mode of development is 'relatively autonomous' (p.17) from the capitalist mode of production which predominates in the world today.

This may seem to be rather strange talk for a Marxist, even for a lapsed one (who still refers to his approach as 'Marxian': Castells, 1994: 19), since what is being suggested is that social change may in fact be determined – to an unspecified degree – by technical advances in production. It is more common for Marxists to argue that it is capitalist principles and practices which shape the direction of change, including technical innovations.

Moreover, Castells' distinction between the informational mode of development and the capitalist mode of production also implies that there is a key realm – technique – which is in some way aloof, or at the least set at a distance, from capitalism, while it is more common for Marxists to argue that capitalism pervades every aspect of society in the present period. In contradistinction to this, in Castells' theorisation there is the assertion that the informational mode of development is autonomous from capitalism, that, therefore, however much capitalism may change, a certain technical realm will therefore remain intact. The logical extension of this is that there must be limits to the changeability of society since the mode of development is, in key respects, beyond the reach of politics. Put vulgarly, 'you may look to a future beyond capitalism, but you're still going to need computer systems to get by'.

Finally, precisely because of this autonomy of the mode of development, we have to ask how much this in itself influences social arrangements. Unavoidably we encounter here the issue of *technological determinism* in the theory of Castells. To the extent that we do, the consequence is a depiction of political activity and aspiration as sharply limited since social arrangements are dependent (if to an unspecified extent) on technical foundations.

What is especially remarkable is just how consonant this theorisation of change is with a great deal of decidedly conservative post-industrial writing. A re-examination of Chapter 3 of this book, the discussion of Daniel Bell, will reveal a strikingly similar approach to that of Castells. Like him, Bell is impressed by the role played in social change by technologies of production. Most recently he has suggested that the 'information society' is established courtesy of ongoing technological innovation which vastly increases the productivity levels of the economy and thereby allows surpluses to be expended on increased service provision. Remember, Professor Bell does not allege that post-industrial society results in the end of political conflicts (hence he refuses the charge of crude technological determinism), it is just that, to him, the prerequisite of everything – even of political disputation – is a technical infrastructure which guarantees productivity at high (if unspecified) levels. Conceptually at least this occupies the same ground as Manuel Castells, as it does that of a host of commonsensical thinking which has it that 'people must eat, drink and find shelter before they start thinking about politics' and that the 'real foundations of society are built on these technical principles'.

The conceptualisations of the new age are remarkably similar in both Bell and Castells. In both thinkers there is the same emphasis on the transformational, indeed foundational, characteristics of changes in techniques of production throughout history and, most recently, in the role of information and knowledge. Thus Castells (1994) refers to a 'technological informational revolution' as 'the backbone (although not the determinant) of all other major structural transformations' (p.20) and he continues to depict an 'informational society' (p.21) which 'replaces the industrial society as the framework of social institutions' (p.19). This is exactly the argument of Daniel Bell which we reviewed earlier in this book.

I mentioned that the roots of Castells' theorisation were in the version of Marxism popularised by Louis Althusser in the early 1970s. It is here that we can identify reasons which explain the conceptual agreement between the Marxian Castells and the explicitly anti-Marxist Bell.

A crucial point to appreciate is that Marxism is an exceedingly diverse body of knowledge, split at the least by one massive fissure. On one side of the Marxist tradition is what has been designated Scientific Marxism (Gouldner, 1980), of which Louis Althusser was arguably the last major intellectual proponent in Europe. This has several characteristic features, but a prominent one has been to draw a distinction between *relations of production* (classes) and *forces of production* (techniques), the former being the realm of class struggle, values and social relationships, the latter that of technology and techniques (factories, machine production, division of labour). There has always been a tension in Scientific Marxist analyses between the relative significance of the forces and relations of production, the former leading thinkers towards stressing the pivotal role of technological development in social change, the latter resulting in an emphasis on radical political action to effect change.

The allegiances and dispositions of most Scientific Marxists led them to argue that the forces and relations of production interacted in such a way that neither was the prime motor of change, but a moment's reflection leads one to realise that a most compelling question revolves round the relative weight of the relationship between the two elements. Put crudely, if one considers that the forces of production are most decisive, then the Marxist scholar will trace change in the development of industrial production; against this, if it is the relations of production from which spring the most decisive changes, then one is drawn towards the revolutionary spirit and organisational capabilities of people. The stock Marxist retort that change is all a matter of the connections between these variables does not help much because, granted that they are each 'relatively autonomous' one from another (a favoured argument of Louis Althusser and like-minded thinkers), then the really vexatious question involves the *degree* of their autonomy.

It took Daniel Bell, in setting out his theory of the evolution of post-industrial society, to articulate a logical extrapolation from this division. What if, he asked, the forces of production turn out to be more significant in social change than the relations of production? What if it is techniques and technologies which divide agricultural, industrial and now post-industrial societies, these being of far more consequence than the relations of production which distinguish between feudalism, capitalism and socialism? In sum, attests Bell (1976a), if 'the forces of production (technology) replace social relations (property) as the major axis of society' (p.80) then 'what has become most important is the emphasis on technique and industrialisation' (p.41).

The entire theory of post-industrialism revolves around Daniel Bell presenting this quasi-(Scientific) Marxist account,[1] insisting that there is a separate yet foundational level of technical production which is of overwhelming import in bringing about social change (Webster and Robins, 1986: 42–48), of such significance that it displaces the relations of production as an important factor in accounting for change.

I would emphasise the common frame of reference employed in Bell's post-industrialism and Castells' informational mode of development. But I would also wish to stress that this was also a frame employed enthusiastically in a particular Marxist tradition that enjoyed considerable popularity in some intellectual circles a generation ago (cf. Thompson, 1978). From his argument in *The Informational City* it appears that Castells has not abandoned all of his allegiance to an Althusserian Marxist approach.

Crucially, it is the identification of a technical/production realm of society which is, in decisive ways, untouched by values, beliefs and social priorities (politics in the most general sense), yet which is simultaneously so axiomatic for society that it is the foundation – and thereby the determinant (though not the only determinant) – of whatever else (choices, institutions, services etc.) is built upon it, that marks the common denominator running through Castells, Bell, Althusser and the tradition of Scientific Marxism.

It may be that there are readers who are puzzled by this commentary, perhaps because they regard the distinction between the realm of the social and that of the technical as self-evidently valid. In reply two points require emphasis. The first is that, if we accept the distinction, then we must also concede that the differences between conservative and radical approaches to understanding contemporary changes are much less remarkable than we might otherwise have presumed. Though the precise degree of emphasis may vary from one thinker to another, the proposition that today the 'information technology revolution' (or the 'informational mode of development') is *the* most portentous factor in bringing about change must be acknowledged. In addition, we must concede that the history of technological innovation is the privileged means of understanding change, since it is from changes in technique that change itself springs.

The second point to underline is that this vision is challenged both from within Marxist traditions and from sociological perspectives. On the opposite side of the Marxist fissure I mentioned above stand Critical Marxists (Gouldner, 1980), amongst whom are assembled the likes of Theodor Adorno, Herbert Marcuse and, in this country, Raymond Williams and Edward Thompson. Here there is a characteristic refusal to privilege technique in examination of social change, either by regarding it as the *primum mobile* of change or by presenting it as something set apart from the wider social world. Instead there is an insistence that technique is part of a whole network of relations under capitalism, relations that have to be understood historically, and in ways which mean that values make themselves felt in the very process of technological development (Aronowitz, 1989). From a different starting point, sociologists concerned with the social construction of science and technology have done much to undermine the faith of those who believe that technologies develop independently of social values (cf. Woolgar, 1988; MacKenzie, 1990; Bijker *et al.* 1987).

The conclusion one needs to reach after this discussion is that Manuel Castells' theoretical premises are at once disputable and at the same time raise profound questions about his attempt to isolate the major factors bringing about social change. As we go on to consider his more empirically grounded observations this theoretical point of departure and its assumptions must be kept in mind.

Having distinguished the capitalist mode of production from the informational mode of development, Castells proceeds towards a more substantive account of recent history. What he discerns here as the most glaring phenomenon is the *crisis* of advanced capitalism that developed in the early 1970s, worsened acutely following the steep rise in oil prices, which has precipitated a sustained attempt at *restructuring* established relations with the intention of revitalising the capitalist economy.

In this the coincidental arrival of the informational mode of development has been critical since it has been harnessed to assist in the energising of capitalist activity. In the words of Henderson and Castells (1987), 'technological progress is being captured within an historically limited economic logic' (p.5). It is the combination of 'restructuring and informationalism' (Castells, 1989: 4) that is the hallmark of recent times: the capitalist mode of production is in crisis and responds by restructuring; the informational mode of development arrives on the scene amidst all of this upheaval. The two are conjoined as capitalist enterprise is rejuvenated by seizing upon the new technologies.

Castells (1989) concedes that capitalism and informational activities have undergone some intermixing, even going so far as to suggest that information processing has been 'fostered' (p.18) by the growth of production and consumption in modern society. Nonetheless, this point is an aside to his major theme: capitalist organisation and the 'informational technological paradigm' (p.17) have developed independently of one another. The former is the domain of politics, choices and values, the latter that of mere – yet foundational – technique.

Rather than repeat at length the objection that this division is far too clear cut, that it drifts easily into technological determinism, and that it too readily presumes that the 'rise of a new technological paradigm' (p.12) occurs without being designed and propelled by social – and even capitalist – values, let us pose the problems by way of an analogy. As the informational cavalry rides to the rescue of a besieged capitalism, how come they knew where to find the Indians? Whence came their weapons, uniforms and provisions? And just who is the most influential player in this scene – the prisoner tied to the totem pole or those infantry with the Gatling guns knocking everything and everyone asunder?

FLOWS OF INFORMATION

Manuel Castells distinguishes the 1945–1973 period as one of relatively stable and prosperous state-regulated, welfarist and corporatist-oriented capitalism. This was turned upside down by recession during the 1970s, and since 1979 we have been moving towards a 'new model of socio-economic organization' (Castells, 1989: 23). This latter, induced by the restructuring to which we have already made reference, is not given a title by Castells, though it is reminiscent of the ideas of those who depict a transition from Fordism to post-Fordism.

Castells emphasises time and again that 'restructuring could never have been accomplished . . . without the unleashing of the technological and organizational potential of informationalism' (Castells, 1989: 29). The new 'socio-technical paradigm' pertains everywhere, and everywhere it presents opportunities for changes which can reinvigorate capitalism. In the workplace increased productivity and new and improved products can come from the application of new technologies to labour and management processes. More widely, 'informationalism' improves the flexibility of production, enabling the introduction of Just-in-Time systems that reduce costs and improve profit margins. Perhaps most importantly, the internationalisation of economic affairs is unthinkable without advanced information technologies.

This line of argument will be familiar to those who have read Chapter 7 of this book. For this reason I do not intend to elaborate it. Instead I aim to focus on some key dimensions of Castells' interest in the interplay between restructuring and the arrival of the informational mode of development. Here emphasis is placed on what he calls the *flows of information*, Castells' core argument being that the development of IT networks around the globe promotes the importance of information flows for economic and social organisation while simultaneously it reduces the significance of particular places. It follows that, in the 'informational economy', a major concern of organisations becomes the management of and response to information flows.

This is in keeping with the geographer's concern with spatial relationships, a central argument being that information networks and the consequent circulation of information results in organisations becoming increasingly able to transcend limitations formerly imposed by place.

We can clarify something of the significance of this by reminding ourselves of the imperative for corporations to restructure to be more competitive over the past twenty years and of the increased globalisation of economic activities that has concurrently taken place. Internationalisation has led to, and been enabled by, the establishment of computer communications networks around the world. In turn, this has helped bring about financial and market integration, with real-time and continuous global transactions. Taken together this has at once stimulated economic recovery by providing opportunities to entrepreneurs and exacerbated the demand to restructure by heightening the risks of competitive failure as markets have increasingly become open to all comers.

Restructuring has taken a good many forms, but an especially pertinent one is that, in a globalised economy, major corporations have been compelled to develop world-wide strategies for production, distribution and sales. Together, internationalisation and restructuring have placed a heightened importance on information flows. Increasingly the priority of capitalist organisations is to manage and process the circulation of information which impinges on market opportunities, investment decisions, labour supplies, component availability, price distributions and production strategies.

Furthermore, it has been the creation of information networks that has allowed decentralisation of many aspects of economic and social organisation to be combined with increased centralisation of decision-making. It is frequently suggested that decentralisation and centralisation are mutually exclusive, but the increasing emphasis on the management of information flows means that a centralised group at headquarters can in fact monitor and co-ordinate highly dispersed (and hence decentralised) organisational interests. Globalised information systems provide corporations with the infrastructure to allow world-wide decentralisation of operations while ensuring that centralised management remains in overall control.

CLASS STRUCTURE IN THE INFORMATIONAL CITY

The flow of information may indeed be emerging as the central feature of the 'Informational Society' (Castells, 1994: 28) and this reliance on networks is reducing the restrictions of place on contemporary activities. Nevertheless, if corporate and financial organisations can increasingly transcend the limitations of place because of information networks, the establishment and operation of these networks have major geographical consequences in themselves. The networks must have nerve centres, places through which the information does not merely flow, but where it is collated, analysed and acted upon. Castells suggests that these nodal points are to be found in certain metropolitan cities which, as they have recently developed, undergo changes in class formation that have major consequences for the conduct of urban life.

Saskia Sassen (1991), looking at New York, London and Tokyo, underwrites Castells' general argument. She identifies a trend towards 'spatially dispersed yet globally integrated organization of economic activity', the result of which is a 'new strategic role for major cities' that function as the 'highly concentrated command posts of the world economy' (p.3). As these 'informational cities' (Castells) have developed in recent years, they have experienced a rapid increase in informational workers whose jobs involve the operation and management of the information networks, thereby providing 'global control capability' (Sassen, 1991: 11). Castells (1989: 184) calculates that upwards of 30 per cent of the workforce of these cities are informational employees situated in occupations ranging from systems analysts, advertisers, brokers and managers to bankers. They enjoyed an especially rapid expansion during the 1970s and 1980s when the infrastructure of the global networks was being constructed at a particularly rapid pace.

This had to be so since corporations need to have headquarters close to stock markets, telecommunications and computing facilities must be ultra-sophisticated, banking and finance organisations need to be located next to major customers, specialist services must be accessible, political interests need to be reachable, and advertisers and media groups need to be proximate to their clients. Of course this is not an iron law, and there will be countervailing influences, but as a rule there are powerful forces impelling concentration of major economic and political players in 'global cities'. Sassen (1991) observes appositely that 'the more global-ized the economy becomes, the higher the agglomeration of central functions in a relatively few sites, that is, the global cities' (p.5). Testament to this is Anthony King's (1990) identification in the mid-1980s of four 'world cities' in each of which are located the headquarters of over twenty transnational corporations (New York with 59, London with 37, Tokyo with 34 and Paris with 26). In turn these draw in the major banks which service them, hence there has been an explosive growth of foreign banks with direct representation in New York and London in recent decades (ibid.: 26).

Together these trends stimulate the growth of information professionals and managers who, controlling and operating the flows of information that are axial to contemporary economic activity, become 'the only truly indispensable components of the system' (Castells, 1989: 30), other workers (including back office functions in finance, insurance and cognate businesses) and/or locations being disposable should circumstances so decree. Saskia Sassen describes these information occu-pations as 'producer services' in order to emphasise their centrality to modern-day economic success since, though they may be far removed personally from practical production, they supply the brains that organise and co-ordinate world-wide economic activity. Employees in these and similar occupations, which Sassen (1991) estimates at between 25 and 33 per cent of the labour force of New York and London, perform the primary task of 'producing and reproducing the organi-zation and management of a global production system and a global marketplace for finance' (p.6).

As we shall see later in this chapter, these professional and technical information experts are found in disproportionate numbers in New York, London, Los Angeles and similar cities (Castells estimates that they constitute 17 per cent of all jobs, but double that in the informational city) not simply because their employers are based there. They also have a way of life which best finds expression in a major metropolitan arena.

Moreover, while information specialists have been expanding in key cities, their significance and visibility have been increased by processes set in motion by the same pressures of restructuring. In this respect uptake of new information technologies has had a particularly marked effect on manual occupations, especially those involved in manufacture. In addition, recent history has seen the frequent removal of industry from metropolitan centres to 'green field' sites, which has exacerbated the trend towards diminution of unionised and skilled workers in urban locales. Illustratively, New York and Greater London lost over half of all manufacturing jobs between the late 1960s and 1989 (Buck *et al.*, 1992: 94).

While the metropolitan cities are undergoing this expansion of information professionals and technicians alongside a reduction in the (relatively) privileged skilled working class, they have also experienced a boom in service occupations. These, however, are not the prestigious producer service jobs (50 per cent of which Professor Sassen [1991: 9] calculates are in the two highest-earning occupational categories); they are more akin to old-fashioned servants of the affluent. Indeed, the vast majority of new service work in the city (which Castells [1989: 184] puts at 75 per cent of all new jobs) is of this kind: low-skilled, low-paid, with low educational demands made of it. These are very often part-time, unstable and casual opportunities to work in the 'informal economy' (Mattera, 1985), where people – most frequently women – are paid in cash to avoid taxation and employment protection legislation, but at the same time have considerable difficulties in getting access to full state benefits. The sort of jobs we are talking about here are office cleaners, waiters, dry-cleaning operatives, bar staff, fast-food joint employees, hotel janitors, dispatch riders, maids, baby minders, shop assistants, and temps of one sort or another. They have expanded in the 'informational city' not least because the information professionals have grown apace, bringing with them high disposable incomes and a lifestyle which requires a plethora of servants. The only groups willing to undertake such work are those incapable of achieving access to the elite information jobs, transients, those unable to escape the city life into which they were born, newly-arrived immigrants without marketable skills, or those driven to the capital when job opportunities in the regions disappeared due to recession and restructuring. Amongst these groups are found a burgeoning *underclass* that needs to be set alongside the vibrant professional managerial class in the informational city.

UNDERCLASS?

The notion of an underclass has caused a very great deal of controversy within and beyond the bounds of the academic discipline of sociology over the past decade or so. A good deal of this stems from the fact that the idea has been media and politics driven, having been taken up by journalists and amplified in books and newspaper articles (e.g. Auletta, 1982; *Times*, 16 May, 1989). I lack space to account for this in detail, but suffice it to say that the concept, at least in certain versions, fits easily with punitive attitudes towards those at the bottom of our society, regarding them as responsible for their demise. Understandably, sociologists (and others) of different persuasions take exception to suggestions that the poor have brought their poverty upon themselves, and with this disagreement comes a ready scepticism about the value of the underclass concept. That said, it is noteworthy that, though it has been pressed most strongly from sociologists of the Right (Murray, 1984), the concept has also been embraced by a wide cross-section of thinkers, including, in the UK, Ralf Dahrendorf (1987), A. H. Halsey (1989), W. G. Runciman (1990) and Labour M. P. Frank Field (1989).

What is common to all conceptions is the attempt to identify a section of society, thought to be about 5 per cent of the total population, which is in some way detached from the rest of the citizenry at the very bottom of society (Pahl, 1988). To some thinkers the underclass is thought to be marginalised not only because it gets minimal economic returns (what work it gets is badly paid, intermittent and uncertain, and it has few prospects; underclass members are heavily dependent on welfare payments for survival), but also because it is trapped in a 'culture of poverty' which leaves it fatalistic and apparently incapable of taking responsibility for itself. In this sense those in the underclass are not just poor, they also manifest a type of poverty that compares badly, say, to the poverty endured by Jewish immigrants to the United States who nevertheless struggled to better their own and especially their children's lives. In contrast, today's underclass are feckless and lacking in initiative, because of which they transmit their unhappy circumstances to their children.

Relatedly, family life seems to be breaking down, at least as evidenced by exceptionally high levels of illegitimacy and never-married single mothers (Dennis and Erdos, 1993), something which testifies to the inadequacy of the underclass at the same time as it condemns children to failure since they lack stable parenting and desirable male role models from whom they might learn (Phillips, 1993). As if this were not enough, the underclass fails educationally through its multiple disadvantages and, as they mature, the fatherless children with poor educational prospects are drawn at worst to the phoney glamour of gang life and drug dealing and, somewhat better, towards the black economy. Thereby they are initiated into a life of criminality and/or petty crime, yet another characteristic of the underclass, 'the habitual criminal [being] the classic member of an underclass' (Murray, 1989: 31). When one adds to this the recurrent suggestion – outright in the United States, *sotto voce* in Britain – that the underclass is disproportionately composed of racial minorities, then it is easy to appreciate that it is an exceedingly controversial notion.

It is not difficult to shoot holes in the underclass concept (Jencks, 1991). Despite all the descriptive details, it has been hard in practice to identify real members of the group (Smith, 1992) and it has proved easy to show that most poor people and the long-term unemployed are neither feckless, nor criminal, nor anti-family (Jencks and Peterson, 1991; Kempson *et al.*, 1994). Further, it does not take very much historical imagination to recall that similar fears have been expressed about the 'undeserving poor', about the fecundity of the pauper classes, about their immorality, depravity, idleness, criminality and general threat to respectable folk (Keating, 1976; Mann, 1992).

Above all perhaps, we ought to be sceptical of 'culturalist' accounts of the underclass which explain it as a result of lifestyle choices while ignoring more robust 'structural' evidence which shows that those at the bottom of society are there largely because there are proportionately fewer jobs around nowadays than twenty years ago, because there is more part-time and casual work available and this translates into lower income going to those who have no choice but to take it, and because government policies since the late 1970s have benefited the rich while exacerbating the conditions of those in poverty.[2]

Whatever one's qualms about the concept, it may be appropriate to describe as an underclass groups that are trapped in the depths of the capital cities about which Manuel Castells writes (Lash and Urry, 1994: 145–70). Here one is particularly drawn towards diagnoses such as those of William Wilson (1987) who, backing away from the underclass concept after excessive media and political appropriation, has had recourse to the term 'ghetto poor' (Wilson, 1991b) to identify those left behind in the inner cities following a haemorrhage of decent-paying and reasonably secure jobs and without means to escape the urban scene to find alternative ways of life. Such groups are conspicuous in large American cities, where they have been augmented by large-scale immigration of desperately poor ethnic minority peoples from other parts of the USA and Southern and Central America especially. Ralf Dahrendorf (1992) lends support to this proposition, observing that 'the American underclass and especially ghetto underclass problem is one of the most serious problems in the entire civilised world' (p.56).

Wilson argues that since about 1970 inner city neighbourhoods have experienced outmigration of working- and middle-class families as well as increased poverty amongst those left behind because of job losses. When he writes that 'the dwindling presence of middle- and working-class households has . . . removed an important social buffer that once deflected the full impact of . . . prolonged high levels of joblessness' (Wilson, 1991a: 461), he is concurring with Castells' (1989) conclusion that what is developing is a city characterised by 'bipolarization' (p.184), a 'dual city' in which we have, cheek-by-jowl, a desperately disadvantaged ghetto underclass and affluent and elite information economy professionals.

The prospect for the 'informational city' is therefore one of increasingly marked social disparities. As the 'global city' consolidates its position as a centre of information management and control, to which end it generates large numbers of professional and technical occupations, so does it create the 'dual city' that is marked by exceedingly sharp class polarisations in which the working class is denuded and the underclass grows. In this way the 'Informational City, the Global City, and the Dual City are closely inter-related' (Castells, 1994: 30).

Not surprisingly, Manuel Castells contends that perhaps the most global city of all, New York, exhibits the most obvious polarisations; it is simultaneously a city of 'dreams and nightmares'. With a 44 per cent share of world capital New York is at once the home of Wall Street and upmarket Manhatten and a city in which some 2,000 people are murdered annually and where the Bronx is synonymous with such acute deprivation that strangers can almost taste the fear when they unknowingly stray into its environ (Fainstein *et al.*, 1992). Wilson (1991b) observes that fully 30 per cent of the American growth in the ghetto poor in recent years is accounted for by New York alone (p.464). Of course the trend towards polarisation is by no means complete and there are other divisions than those between information professionals and the underclass (cf. Mollenkopf and Castells, 1991). Nonetheless, the contrast between a 'metropolitan heaven' and an 'inner city hell' (p.407) is palpable.

Castells' general thesis requires qualification (Marcuse, 1989), especially when we get to the level of individual cities. For instance, it is clear that the three leading world cities (with 80 per cent of world market capitalisation: Castells, 1989: 340–341), New York, London and Tokyo, have significantly different characteristics, Tokyo being considerably more homogeneous and settled (Sassen, 1991: 307–315) than either London or New York. However, I do believe that one can readily appreciate the force of Castells' case. Let us look at a major city which Castells himself considers displays the stratification features of the informational city.

Los Angeles is undisputedly a global city and it has a well documented social ecology. Castells (1989) observes that it is especially since 1970 that LA has become 'a leading financial and business center at the international level' (p.218), being the major hub for trade and investment in the Pacific Basin. In line with this Anthony King (1990) reports that by 1988 70 per cent of downtown LA was owned by foreign capital. An upshot has been sustained growth in professional and managerial jobs that have in turn stimulated a huge demand for servant-type occupations. This has been met, in large part, by a continuous stream of immigrants, legal and illegal, to Southern California.[3] Alongside this LA has suffered job losses in manufacture (Castells, 1989: 219), not least in defence-related businesses which have been the backbone of California's industrial structure and are now in serious trouble following the end of the Cold War.

Writer David Rieff (1991) was led by what he saw in his sojourn in the city to title his arresting book *Los Angeles: Capital of the Third World*. He came to this conclusion because he was so struck by the *close physical proximity* of the affluent professionals and the underclass (East Los Angeles and the Watts district are but a few minutes on the freeway from the opulent homes on the west side of the city), the *interdependence* of the two classes (the rich for services rendered, the poor for survival incomes), and the *yawning chasm* that separated their lives.

Illustrations of this are easy to find. On the one hand, maids are an essential element of the professionals' lifestyles, to cook, to clean, to look after children, to prepare for the dinner parties held in the gaps found in the frenetic work schedules of those deep into careers in law, corporate affairs, trading and brokerage. The maids, generally Hispanics, ride the infamously inadequate public transit buses to points in the city where their employers may pick them up in their cars to bring them home to clean up breakfast and take the children off to school. On the other hand, visitors are often struck by how verdant are the gardens of those living in the select areas of LA. Often they make the assumption that 'anything grows here in this wonderful sunshine'. But they are wrong: Los Angeles is a desert and gardens need most intensive care to bloom. They get it from an army of mainly Chicano labourers which arrives on the back of trucks very early in the mornings to weed, water and hoe – for a few dollars in wages, cash in hand.

In spite of this dependence, which obviously involves a good deal of personal interaction, the lives of the two groups are very far apart. Of course this is largely because they occupy markedly different territories, with members of the poor venturing out only to service the affluent on their terms as waiters, valets, shop assistants and the like. The underclass also inhabit areas which the well-to-do have no reason (or desire) to visit.

To Mike Davis (1990) this territorial segmentation is evidence of an 'increasing South Africanization of its [LA's] spatial relation' (pp.227–228). Such terminology is scarcely an exaggeration: as one crosses the city on the freeways in one's air-conditioned car the only sight one might get of the urban ghetto is a glimpse of boarded-up housing and occasional groups of men standing around on street corners, seen from a comforting distance on the Santa Monica Freeway that is suspended above the streets below. It seems literally *like another world.*[4]

As Rieff (1991) again points out, circumstances are such in Los Angeles that the well-off have no need to encounter the poverty in their own backyards. LA is a 'famously pleasant place to visit. Its horizontality, and the absence of any efficient and fast system of public transportation, means that unless one is poor or chooses to go to those parts of the city where the poor live, it is easy to forget that anyone in LA isn't middle class' (pp.29–30).

The two sides of the dual city may not come into contact with one another except on pre-determined terms, but the affluent are well aware of the areas in which they must not stop their cars. The territories occupied by the underclass are ones of serious crime, of gangs (chiefly black and Hispanic), drug-dealing, everyday murder and maiming, and grave dangers to the innocent or unprepared intruder. For these reasons they are rarely entered by outsiders who are quickly attuned to the prohibitions of the streets. Public spaces such as parks and monuments become diminished, places to be avoided due to the presence of the underclass.

Another side of what Davis (1990) has called 'Fortress LA' is the elaborate security – tantamount to a 'militarization of city life' (p.223) – which the successful mount around their homes, with video cameras, electronic systems, 'armed response' warnings and hired guards touring neighbourhoods to keep out marauders.

As long ago as 1973 Tony Giddens, introducing the concept into his class analysis of advanced societies, noted that 'we may undoubtedly expect chronic "hostile outbursts" on the part of members of the underclass' (1981: 218). Given the polarisation of classes we ought not perhaps to be surprised that, for several days beginning on 29 April 1992, LA experienced serious rioting which left scores dead, thousands injured and millions of dollars of property burned and looted. This was not 'class war' in that it was not directed against the well-to-do (Korean shop owners in the poor localities took the brunt of resentful attacks), but its ferocity is a portent of what the 'dual city' might be in the future.

The trigger was the on-camera beating of a black man, Rodney King, by Los Angeles Police Department officers, but who can doubt the underlying cause? An *Observer* reporter (Stephen, 1992) articulated it thus:

It is no coincidence . . . that it [rioting] all happened in the American city that most epitomises the burgeoning growth . . . of a powerless underclass – a Rich v Poor polarisation in a city where the world's most obscene conspicuous consumption of wealth exists so closely alongside Third World-type ghettoes, where Bel Air can seem like exclusive parts of Johannesburg and South Central Los Angeles more like Soweto.

To be sure, urban divisions are not new (Glasgow, 1980), but these now appear to be forming in the informational city in especially vivid colours. As Castells says, a 'new form of urban dualism is on the rise, one specifically linked to the restructuring process and the expansion of the informational economy' (1989: 224).

There is something else about the experiences of these divided cities which deserves notice. The polarisation is undeniably dramatic: the TV nightly news is full of horror stories from the East and Central sides of the town, stories of apparently random killings, of rapes, gang battles, homelessness, addiction, of despair and hopelessness, of all the desperate ills that are visited upon the urban poor. Yet, though all this may be but a few blocks away, and though it enters one's home in dramatic form on the television console, it is still far removed from one's direct experience. Rather it seems just another part of the television coverage, as real, but little more so, than that coming from Beirut, Belfast or Belgrade. Rieff (1991) recalls that

On more occasions than I like to remember, I would sit in some Westside living room, surrounded by lovely objects and by people with soft voices, and watch video clips on the day's mayhem in Compton or Boyle Heights. Though I would will myself to believe in the reality of these images, I often found myself only half doing so. The comfort around me was too strong, the horrors on the screen too removed, too weightless. They might as well have been taking place on the moon.

(p.39)

Those who most benefit from the informational city seem to be able to find ways in which they can live in close proximity to the ghetto underclass, even to employ many of its members to service their needs, but they are also able to develop ways of keeping them and their problems at an emotional as well as at a physical distance. To examine this further we might look at the changing culture of these cities.

THE POSTMODERN CITY?

Castells (1989) observes, almost *en passant*, that though the 'information producers' which form a 'new professional-managerial class' are a numerical minority, nevertheless they dominate the city culturally. They are, in his terms, a 'hegemonic social class' that might 'not necessarily rule the state but fundamentally shapes civil society' (p.228). Castells does not himself explore much of the cultural implications of the informational city, but in so far as these give expression to the ways of life of its inhabitants then they are worth serious consideration. We know from Castells and others that the informational city owes much of its present character to the pressures that stem from managing globalised information flows, but what about the informational dimensions of this new urban form? Just what way of life, what cultural representations and features, do we come across in the informational city? What follows is somewhat speculative, but readers might want to reflect on the proposition that the informational city is also the *postmodern city*.

We do know that the information professionals and managers have considerable power that springs from their high levels of educational attainment and earnings power. We know too that 'they are attracted to the amenities and lifestyles that large urban centers can offer and are likely to live in central areas rather than in suburbs' (Sassen, 1991: 12). An interesting set of questions revolves around the lifestyles that these yuppies (as the young, upwardly-mobile professionals were caricatured in the 1980s) exhibit and how city life is constructed to accommodate their preferences.

At the outset we may contrast their orientation with that of members of the underclass. Here what is most striking is the global outlook of the information professionals which is opposed to the intense localism of the underclass. We ought not to be surprised at the 'cosmopolitanism of the new informational producers' (Castells, 1989: 227). Their everyday life is a matter of moving along computer-communications networks that traverse the globe, their constituency is international, their employers frequently have interests in dozens of countries, many of them are business travellers as a matter of routine, their incomes (and pleasures) impel them towards exotic restaurants and cuisines, they interact regularly with colleagues and clients from a variety of nations, their mind set is one which thinks globally and which is 'open to messages and experiences that embrace the entire world' (ibid.).

Conversely, it is the localism of the underclass that most impresses. Members here are frequently locked into a familiar and fixed pattern of relationships, into a neighbourhood with distinct territories, and they often lack the means or motives to travel any distance from their own realm. It is astonishing that, amidst the 'global city', there are segments of the population that rarely leave it, for example never moving from Stoke Newington to visit say Buckingham.

Another dimension of this divide is the differential access to information the groups experience. The upper echelons of the stratification system both know how to access and have the wherewithal to get information from many sources in many locations: the data base is ever-ready, the telephone no problem, the facsimile machine routinely used, the network freely roamed, global news sources to hand. Down in the underclass things are very different. The poor tend to 'shrink the world to their specific culture and their local experience, penetrated only by standardized television images, and mythically connected, in the case of immigrants, to tales of the homeland' (Castells, 1989: 227). With such a contrast we may legitimately talk here of the information rich and the information poor in the world city. We may remind ourselves too that they might have some contact with one another, but, as David Rieff (1991: 102) notes, while the cosmopolitan professionals converse knowledgeably at dinner parties about German unification and the turmoil in Bosnia, they remain apparently oblivious to the lives and experiences of their maids back in the kitchen.

But the culture of the informational city is much more than a matter of access and orientation to sources of information. The restructuring of capitalism, in bringing to the fore professional and technical experts working in the City, advertising, financial services, banking, corporate headquarters, accountancy firms and software houses, has made a very much greater impact on the urban way of life.

David Harvey (1988), in a short but insightful article, argues that this results in 'voodoo cities', evoking here a comparison with a carnival mask that is gaudy, brash and colourful, while behind the 'fun' disguise is something rotten. This theme has it that the cultural upheaval that is taking place and which is highly visible and up-front (bright and cheerful clothes, colourful and appealing shops, attention-grabbing architecture, a cult of pleasure, an enterprising 'can-do' ethos) disguises the decline of many other important aspects of city life, notably the working class and its occupational communities.

The redevelopment of London's Docklands is a prime instance of this process. London's docks virtually all closed between the late 1960s and the early 1980s (taking with them thousands of manual jobs that had sustained communities in the East End), at the same time as London was promoting its role as a global information centre. The City itself had to find room to expand in response to this pressure, and it did so, creating additional office space in parts of the West End (noticeably around Covent Garden) and, most strikingly, opening branches to the east in Docklands. The Canary Wharf project, aiming to provide 71 acres and 50,000 jobs, was the most ambitious attempt to use the former docks for offices, expensive accommodation (close to the office, but unsuitable for children, hence ideal for yuppies), state-of-the-art rail links to the City, high-class restaurants, and an appealing ambience designed with the informational professionals uppermost in mind. The project encountered some serious financial problems in the early 1990s and began to stall, but a good deal of it had already been constructed. Those living and working in the area beforehand, the London working class, had been pushed aside, with the inevitable result being a 'growing polarization' of classes and an uneasy co-existence of new clusters of offices and upmarket housing sited beside 'concentrations of working-class households, many now unemployed, living in deteriorating high-rise, high-density public housing' (Harloe *et al.*, 1992: 187).

Moreover, changes taking place increasingly *exclude the poor* by, for example, a marked expansion of housing and specialist estates which are gated and guarded to keep out the 'dangerous classes'. This is evident in the above comment on the Docklands redevelopment, but it is much more widespread than that. In and around Los Angeles the search for 'absolute security' has led to the creation of 'high-tech castles' (Davis, 1990: 248), there seeming to be an overwhelming pressure to fortify, with electronic devices and/or 24–hour-a-day security, against the 'lawless'.

Separation goes further even than designing homes to keep at bay the 'criminal classes'. It extends also to public spaces, especially to newly created shopping areas (often outside the city centre, where totally new construction allows maximum exclusion of undesirables by, for example, making the shopping mall a sealed space, to enter which one must pass under the gaze of security guards and electronic monitors) and refurbished older shops, office areas and even precincts. In these latter the installation of video cameras is now routine, deliberately meant to record all that happens so that miscreants are likely to be apprehended.

As Mike Featherstone (1991) puts it, it is now a 'central principle . . . that these are privately owned public spaces in which the public are under the watchful eye of video-cameras, and rowdy, troublesome elements are excluded before the disorder might disturb others' (p.105). Mike Davis (1990) goes further, to designate Los Angeles a 'carceral city' and its shopping centres the 'panopticon mall' in which consumers may feel safe precisely because potential threats are under permanent surveillance.

If the informational city is characterised by a culture in which the built environment is intended to keep at bay the poor and to secure the safety of the affluent, then there is another feature which is still more distinctive of its way of life. This is that, more than ever, the informational city shifts towards being a 'landscape of consumption' (Zukin, 1991b: 37).

This in no way contradicts my earlier remarks on the centrality of producer services to the global city. Cities such as Los Angeles and New York are central to the continuing success of advanced capitalism in that they are the informational hubs of world-wide production activities. However, while organised globally through these cities, production itself has been largely evacuated from such places, leaving them as prime locations of consumption.

Sharon Zukin (1991b) emphasises that globalisation has brought about an increased disconnection of *market* and *place*. The importance of place diminishes as production and marketing have become global, leaving some modern cities as shrines to consumption of products that come from pretty well anywhere in the world. This creates difficulties for cities since places strive for stability and constancy of meaning, to which end cities may go to great lengths to develop appropriate (often tourist-oriented) images, but this itself is subverted by the constant changes which the global market induces in the consumption-oriented city.

This instability of life in the informational city, something constantly stimulated by innovations in the realm of culture, is thought to be a defining feature of the postmodern experience. Nothing is fixed, nothing is agreed, no meanings last . . . . The postmodern city is a maelstrom of change: in fashions, in opinions, in symbols, in designs, even in architecture.

At the same time, the changing class composition of the informational city has meant that a great deal of the settled working class has been forced out of town and its established culture of extended family, community and occupation destroyed. Meanwhile informational professionals have moved into 'gentrified' and newly-built parts of the city. They live here at once because they can be located close to their offices and because it is the postmodern city's culture of change, uprootedness, variety and fragmentation which especially appeals to them (Zukin, 1991a: 247).

This generation of professional and managerial workers, who came of age during the 1960s when modernism was coming under sharp attack, has played a principal role in the reconstruction of city life. It was this group, with its enthusiasm for recuperating 'history' (in Camden Town, in Spitalfields and other parts of London) that bought artisanal housing on the cheap and restored chosen and distinctive features. It was this class, with its high spending power and enviable educational achievements, that was drawn towards and further stimulated urban culture (Zukin, 1988), that, in the view of David Harvey (1989a), instituted 'the turn to postmodernist styles' (p.272).

There are many aspects of the informational city that one might seize upon to illustrate its postmodern character. High amongst them would be a discernible breakdown of what one might consider to be 'traditional' or 'modern' behaviours, dress styles and even moralities. For instance, orthodox business suits, still more pin-stripes and bowler hats, have given way to bold and bright colours, loud braces and designer shoes, while the upper-class accents and received pronunciation is now interspersed even in the City of London with a 'barrow boy' vernacular. There has come about a related scepticism as regards the reasons once used to justify one's work and earnings. Where once perhaps the business community in London would justify their jobs in terms of a creed such as 'for the good of the family', 'loyalty to the company' or even a 'Christian ethic', now the notion of a guiding set of principles of any sort is abandoned, people working for nothing more (or less) than money and what it might buy them. There has emerged an associated brashness, an 'if-you've-got-it-then-flaunt-it' outlook, which stands in marked contrast to the modernist practice of 'gentlemanly decorum'.

Relatedly there has come about an appreciable narrowing of the distinctions between high and low culture and a loosening of the class and educational relationships between them. Where perhaps there was formerly a close connection between profession, university background and aesthetic preferences (for example, law, Oxbridge, ballet and live theatre), this has significantly weakened. At the least the assumptions of superiority that were vested in such ways of life have come under attack from a determined relativist position which sees culture as no more and no less than a matter of *lifestyle choices*. With this has come a willingness to mix sharply different cultural forms, enjoying the opera as well as the rock concert, enthusing about Arsenal football club while subscribing to the London Library.

Further, enjoyment is to the fore of the postmodern city. And this is a hedonism without scruples, without a guilty conscience. The tyranny of meaning, from the anguished 'why am I doing this?' to the searching 'what is the real message of this play?', is abandoned, to be substituted by the primacy of pleasure as a motive force. This is, of course, to evoke a major theme of postmodernism, but it is arguably in the informational city, amongst the professional and managerial classes of the younger generations, that this finds its most distinctive form. Functionality of architecture as well as deep motives that might guide actions, have given way to a pursuit of pleasure in all things. A key element of this is an aestheticisation of life (Featherstone, 1992), seen everywhere from the design of office blocks (built not for efficiency but because they look outrageous and memorable), to an emphasis on style in dress, all to indicate a culture of fun, excitement and sensory enjoyment rather than dutiful fulfilment of the job because that is the proper thing to do.

An emphasis here is on the simulational character of contemporary life – and there is nothing more artificial than city life – and with this stress all notions of the authentic are abandoned. The postmodern urbanites, at least those with the where-withal to afford it, have access to all manner of cultural forms which have been brought to the global city, each and any of which can be experienced solely for the pleasures they might bring, with no hangups about whether or not they are authentic. Exotic cuisines may thrive, musics from diverse nations, cultures and sub-cultures can be explored, ethnic clothing can be eclectically enjoyed (Cooke, 1988).

At the heart of all this is consumption, and, perhaps most notably, *shopping*, which in the postmodern city takes on a primary cultural role. Now cities have long been centres of consumption, with, for example, department stores being a prominent feature for a century or more. However, one should acknowledge the shifts that have occurred over recent decades, away from what one might consider an instrumental approach to shopping ('I go to the shop for a pound of butter'), towards shopping as an aesthetic experience. Here we are referring to shopping as *an end in itself*, as a pleasurable experience that is heightened by the freedom to roam amongst what is on offer without being approached by a salesperson, by the ease of access to open-doored and inviting shops, by the heterogeneity of boutiques and specialist arcades, by the Musak, sumptuous carpeting and pleasant scents that enhance mood and sensations, by the carefully planned lighting, colour co-ordination and displays of goods. . . . There is a slogan which captures this well (and in appropriate parodic form): 'I shop therefore I am'.

Finally, we might return to the issue of polarisation in the informational city to ponder the question: how do the successful manage psychologically to live with themselves when they are so close to everyday reminders of the underprivileged? We know that they are well aware of dangerous territories which are to be penetrated only with caution, but how do they accommodate themselves to the acute poverty and decrepitude that surrounds them? To be sure, cities have long been divided and the relatively privileged have always managed to find explanations for this which have left them free to enjoy their own favoured circumstances. Generally speaking this involves one form or another of blaming the victim ('they drink, they're feckless, they're idle . . . '). And yet, even at the height of the Victorian epoch there were scruples, the nagging conscience of the Christian gentlefolk when confronted by child labourers, the homeless, and the maltreated.

No doubt some of this remains,[5] yet at a time of sharpening urban differences –when, for instance, it is impossible to visit London without coming across the dispossessed and when it is hard to plead ignorance about such inequalities since they regularly feature on television programmes and in newspaper articles – how is it that those quite directly involved, the urban professionals, can apparently adjust to this situation?

It is hard to resist the conclusion that postmodern orientations help. If conceptions of the 'true', the 'authentic' and the 'real' have been undermined, then so too has the idea of an integrated self, a 'real me' that is guided by a coherent creed, principles and morality. In so far as postmodernism is about the primacy of experience over meaning, about the prioritisation of the *decentred* self over modernist notions of an authentic individual, then this must help explain how one may feel a lack of shame (as a dictum of the Eighties pronounced: 'Guilt is for losers!') about one's own prosperity and pleasures while living close by others whose lives are squalid and blighted.

CONCLUSION

Manuel Castells' focus on the importance of information flows, and amidst these the formation of global cities, has been the subject of this chapter. The establishment of information networks is undeniably a defining feature of the late twentieth century and Castells' analysis of the relations between these networks and changes in the urban ecology is arresting. The *informational city* is a major contribution to our thinking about the significance of information in the world today.

However, Castells' theoretical starting point, his reliance on the concept of an 'informational mode of development', is disconcerting. It is so because it puts him, Marxian language notwithstanding, in common accord with a good deal of conservative thinking about the role and significance of information nowadays. At the same time, it easily drifts into a form of technological determinism found most frequently amongst techno-boosters who insist that the 'information revolution' will transform the way we live.

This criticism apart, Castells' substantive analysis of the informational city is illuminating, particularly his identification of changed class relationships and an associated process of polarisation. Certainly one might want to debate the degree to which all global cities conform to this pattern as well as precisely which cities merit the designation (Paris? Chicago? Frankfurt? Hong Kong?–Markusen and Gwiasda, 1994). I suspect too that historians might want to take issue with Castells on the novelty of the divided city. For example, a century or so ago 'outcast London' (Jones, 1984) was very much feared by respectable opinion, the 'casual poor' being just as big as today's ghetto underclass and just as often seen as mendicant, dissolute and criminal. These 'dangerous classes', crammed into the rookeries (ghettoes in today's parlance) in the East End, were widely regarded as alien and threatening to the the affluent who moved out of the vicinity to create a *cordon sanitaire* between themselves and yesterday's underclass (Mayhew,1971).

Nevertheless, whatever qualification one might wish to make, there is a considerable body of evidence, notably on Los Angeles, London and New York, which does lend support to Professor Castells' thesis on the informational city. Globalisation and information networks are encouraging a reformation of class relationships in certain cities and the divisions arising from this are palpable.

Finally, my observations on the postmodern characteristics of the informational city are speculative and go further than Manuel Castells would allow. I am neither confident of the empirical evidence about the postmodern city nor about the generally glum reasoning that informs my account. After all, the suggestion that postmodernism is lacking in substance, in coherence, and even in morality, need not be interpreted in a condemnatory way. Quite the contrary, this very superficiality and changeability of the city can be regarded in a positive light (Raban, 1974). In the words of Richard Sennett (1970), 'the jungle of the city, its vastness and loneliness, has a positive human value' (p. xvii), its very disorder, complexities and challenges encouraging people to grow out of an adolescent yearning for a conformist 'community' (of suburbia or a rural idyll that never existed) and also to get beyond the planned and prescribed arrangements characteristic of modernism. I am not persuaded of this myself, and am content to identify a consonance between the lifestyles of professional and managerial workers and postmodern attitudes.

10

CONCLUSION

The purpose of this book has been to examine the significance of information in the late twentieth century. It has asked how and why is it that, on the brink of the second millennium, information has come to be perceived as a, arguably the, defining feature of our times. My starting point was to remark on the consensus amongst thinkers that information is now of pivotal importance in contemporary affairs: it is acknowledged that not only is there a very great deal more information about than ever before, but also that it plays a central and strategic role in pretty well everything we do, from business transactions and leisure pursuits to government activities.

But beyond these observations consensus about information breaks down. While everyone agrees that there is more information and that this has increased in pertinence nowadays, thereafter all is disputation and disagreement. Recognising this, I have tried to identify the major attempts to understand and explain what is happening in the information domain and why things are developing as they are, at once to make clear the bases of different approaches while simultaneously testing them against available empirical evidence, against one another, and with any additional critical insight I could muster.

It must be in the detail of this exposition and assessment of varying 'theories of the information society' that the value of this book is to be found. So much commentary on the 'information age' starts from a naïve and taken-for-granted position: 'there has been an "information technology revolution", this will have and is having profound social consequences, here are the sort of *impacts* one may anticipate and which may already have been evidenced'. This sets out with such a self-evidently firm sense of direction, and it follows such a neat linear logic – technological innovation results in social change – that it is almost a pity to announce that it is simply the wrong point of departure for those embarking on a journey to see where informational trends, technological and other, are leading. At the least, recognition of the contribution of social theory moves one away from the technological determinism which tends to dominate a great deal of consideration of the issues (though, as I restate below, with some social scientist thinkers subtle – and sometimes not so subtle – technological determinism continues to linger).

More than this, however, I do think that one's appreciation of the significance of information in contemporary life is immensely deepened by encounters with the likes of Herbert Schiller and Anthony Giddens, or with Regulation School and postmodern thought. Who cannot be stimulated, for example, by Daniel Bell's arguments that it is the increase in service employment which leads to an expansion of information occupations that have most important consequences for how 'post-industrial' societies conduct themselves? Who cannot find arresting Tony Giddens' contention that the origins of today's 'information societies' are to be found in surveillance activities that are largely driven by the exigencies of a world organised into nation states? Who cannot take seriously Herbert Schiller's suggestion that the information explosion of the post-war years is the consequence, for the most part, of corporate capitalism's inexorable march? Who is not provoked by Jürgen Habermas' fear that the 'public sphere', so essential to the proper conduct of democracies and where the quality of information supplies the oxygen which determines the health of participants, is being diminished? Who would not concede the relevance to understanding information of theorists of a purported transition from Fordist to post-Fordist forms of socio-economic organisation? Who cannot be intrigued by Jean Baudrillard's gnomic observations on signs that are simula-tions or Jean-François Lyotard's identification of a 'principle of performativity' underpinning the generation and application of information in the 'postmodern' era? Who cannot learn from Manuel Castells' insights into the development and dangers of the 'informational city'? And who, encountering these thinkers and the calibre of their work, cannot but conclude that most discussions of the 'information age' are hopelessly gauche?

Of course it would be disingenuous of me to stop here with the suggestion that all I have tried to do is introduce readers to a variety of interpretations of informa-tional trends. Those who have gone this far in the book will have realised that I have found certain thinkers more persuasive than others. I have endeavoured to make this, and the reasons why I favour them, clear as I have gone along, but by way of a conclusion a few summative observations may be in order.

I believe that if one is trying to make sense of the information realm and its import in the present age, one should be drawn primarily towards the ideas and research, above all, of Herbert Schiller, Jürgen Habermas and Anthony Giddens, as well as to the significant body of work that has been influenced by their themes. This does not for a moment mean that the contributions of Daniel Bell, or of Manuel Castells, or of Mark Poster and other scholars are negligible, still less worthless. Quite the contrary, I have attempted, when analysing such thinkers, to indicate and evaluate the positive elements of their work as well as to point out any weaknesses I may have found in it.

Nevertheless, there are two major reasons for my preferences for some thinkers rather than for others. The first concerns the capacity of these approaches to illuminate what is actually going on in the world and how well their propositions stand up to empirical scrutiny. On the whole the Critical Theory of Herbert Schiller (in whose writing the theory is decidedly and advantageously subordinated to a concern with substantive developments) and Jürgen Habermas, and the historical sociology of Anthony Giddens, seem to me more persuasive than the writings of post-industrial and postmodern enthusiasts.

Perhaps to state the obvious, to admit my preferences means neither that I endorse everything each of these scholars forwards nor that Schiller, Habermas and Giddens are altogether agreed on the salient features of the informational domain. It will be obvious to readers that Schiller's focus on the imperatives imposed by capitalism differs from Habermas' concern with the requisites of democratic debate, and both differ from Giddens' emphasis on ways in which the state especially, and particularly in its military and citizenship dimensions, influences the collection and use of information.

However, there is one crucial point of agreement within the diversity of views of these thinkers and it is something which sets them apart from those other contributions that I have found less helpful in understanding and explaining the role of information in contemporary affairs. This takes me to the second reason for my preferences. What Schiller, Habermas and Giddens do share is a conviction that we should conceive of the *informatisation* of life, a process that has been ongoing, arguably for several centuries, but which certainly accelerated with the development of industrial capitalism and the consolidation of the nation state in the nineteenth century and which has moved into overdrive in the late twentieth century as globalisation and the spread of transnational organisations especially have led to the incorporation of hitherto untouched realms – far apart geographically and close to areas of one's intimate life – into the world market.

These scholars believe that informational developments must be accounted for in terms of historical antecedents and *continuities*. Each of them therefore prioritises in their separate accounts phenomena which, over time, have shaped, and in turn have built upon, informational patterns and processes to ensure, as best they could in uncertain and always contingent circumstances, that existent social forms might be perpetuated. For instance, in Herbert Schiller's work we find a recurrent insistence that it is capitalist characteristics which predominate in the origination and current conduct of the informational realm: it is the primacy of corporate players, of market principles and inequalities of power which are most telling. Similarly, those who argue that the 'public sphere' is being diminished have recourse to explaining the expansion of misinformation, disinformation, info-entertainment – information management in all of its guises – in terms of the historical expansion and intrusion into all spheres of life of commodification and market criteria. The 'information explosion' is to these thinkers comprehensible as an integral part of the up and down history of capital's aggrandisement.

Again, Giddens' approach to information places its development in the context of the development of nation states and associated historical patterns of the making of modernity such as the industrialisation of war and the spread of citizenship rights and obligations. A similar emphasis comes from Regulation School theorists who explain informational trends in terms of requisites and outcomes of advanced capitalism following recession and restructuring brought about by the threats and opportunities associated with the spread of globalisation.

It must be stressed that those who emphasise historical continuities are not alleging that nothing has changed. Quite the reverse: the very fact of *informatisation* is testament to their concern to acknowledge the changes that have taken place and that these are such as to promote information to a more central stage than previously. Nevertheless, what they do reject is any suggestion that the 'information revolution' has overturned everything that went before, that it signals a radically other sort of social order than we have hitherto experienced. On the contrary, when these thinkers come to explain informatisation they insist that it is primarily an outcome and expression of established and continuing relations. It is the conviction of each of these thinkers that the forces they have identified as leading to the informatisation of life still prevail as we approach the second millennium.

My reason for preferring the idea of an informatisation of life which stems from the continuity of established forces become clearer when we contrast it with the propositions of the likes of Daniel Bell, Larry Hirschhorn, Gianni Vattimo and Mark Poster. Here, again amidst marked divergences of opinion and approach, is a common endorsement of the primacy of *change* over continuity. In these approaches change is regarded as of such consequence that reference is recurrently made to the emergence of a novel form of society, one which marks a *system break* with all that has gone before. Various terms are used by such thinkers, from the generic 'information society' to 'post-industrial society', 'postmodernism', and 'flexible specialisation'.

Of course none of these thinkers is devoid of historical imagination, but the emphasis of their analyses constantly centres on the novelty of the 'information society', the fact that it is set apart from anything that has gone before. I have tried to demonstrate throughout this book how this proposal is unsustainable and while making my case I have found myself returning time and again to those who argue for the primacy of continuity.

In addition, it is noticeable that so many of those who argue for the emergence of an 'information society' use deterministic explanations for the coming of the new age. These are certainly considerably more sophisticated than the crude technological determinism adopted so enthusiastically by techno-boosters such as Alvin Toffler (1990) and James Martin (1978). Nonetheless, there remains a strong undercurrent of technological determinism in those who conceive of a 'second industrial divide' (Piore and Sabel), a new 'mode of information' (Poster) or an 'informational mode of development' (Castells). Moreover, as Krishan Kumar (1978) definitively showed several years ago, behind Daniel Bell's concept of post-industrialism lies a similarly, if more sophisticated, deterministic account of change, this time through the hidden hand of 'rationalisation' which, of course, finds its major expression in the application of improved technologies but which also is evidenced in the development of more refined organisational techniques.

In this book I have been at pains to underline the shared way of seeing of thinkers who, however apart they might seem at first sight, hold in common certain principles. For those who assert that we are witnessing the emergence of an 'information society', high on their list of shared principles is technological (or in Bell's case technical) determinism.

To repeat the two major complaints about such an approach: it at once singles out technology/technique as the *primum mobile* of change (which is oversimplistic) while simultaneously presuming that this technology/technique is aloof from the realm of values and beliefs. I do not think it has been difficult to demonstrate that this is a misleading perception, but it will keep infecting analyses of informational developments. Above all, it seems to me, it is an approach which misconceives social change because it desocialises key elements of social change, persistently separating technology/technique from the social world (where values and beliefs are found), only to reinsert it by asserting that this autonomous force is the privileged mechanism for bringing about change. Not surprisingly, those who envisage a dramatic but asocial 'information technology revolution' and/or radical shifts in technical efficiency, are easily persuaded that these *impact* in such a manner as to bring about an entirely novel form of society.

As I argued in Chapter 2, those who contend that an 'information society' has arrived (or is in process of arriving) in recent years operate with measures that are consonant with this technical determinism. It is striking that they seek to identify the 'information society' by counting phenomena which they assume characterise the new order. These may be information technologies, the economic worth of information, the increase in information occupations, the spread of information networks, or simply the obviousness of an explosive growth in signs and signification (which therefore do not need to be counted).

Subscribers to the notion of an 'information society' quantify some or other of these indicators and then, without justification, claim that these quantifiable elements signal a qualitative transformation – the emergence of an 'information society'.

Similarly, when one presses forward to examine their definition of information itself, most often we come across a related principle: information is presumed to be a quantifiable phenomenon that is separable from its content – it is so many 'bits', or so much 'price', or so many 'signs', seemingly anything but something which has a meaning (though, as Theodore Roszak, 1986, eloquently reminds us, to most people the content of information – what it means – is of the essence). Then, having adopted a non-semantic definition of information that can more readily be quantified, we again come across the allegation that a quantifiable increase in information heralds a qualitative change in society and social arrangements (an 'information society').

It appears to me that those who explain informatisation in terms of historical continuities give us a better way of understanding information in the world today. This is not least because they refuse to start with abstract measures of the 'information society' and of information itself. While of course they acknowledge that there has been an enormous quantitative increase in information technologies, in information in circulation, in information networks and what not, such thinkers turn away from such asocial and deracinated concepts and back to the real world. And it is there, in the ruck of history, that they are able to locate an information explosion that means something substantive and that has discernible origins and contexts: that *these* types of information, for *those* purposes, for *those* sorts of groups, with *those* sorts of interests are developing.

NOTES

2 INFORMATION AND THE IDEA OF AN INFORMATION SOCIETY

1   In the economics literature 'information' and 'knowledge' tend to be used interchangeably. Sometimes, however, knowledge is used in the sense of an aggregation of different
types of information, thus allowing economists to consider ways in which it has
properties of a 'stock' and 'flow' (Melody, 1987: 314).

3 THE INFORMATION SOCIETY AS POST-INDUSTRIALISM: DANIEL BELL

1   Bell (1979) distinguishes the terms conceptually as follows: information means 'data
processing in the broadest sense'; knowledge means 'an organised set of statements of
fact or ideas, presenting a reasoned judgement or an experimental result, which is
transmitted to others through some communication medium in some systematic form'
(p.168). In practice he often uses the two terms interchangeable.
2   A classic article by John Goldthorpe identified over twenty years ago a 'recrudescence
of historicism' amongst social scientists, with Bell explicitly included in those critiqued.
Goldthorpe pointedly observed that the thinking of these social scientists rested upon
historicist assumptions 'even though historicist arguments may not be openly advanced
or may be actually disavowed' (1971: 263).
3   This is highly contestable, and of course hinges on one's definition of 'professional'.
Against Bell one may note that the Office of Population Censuses and Surveys in 1990
categorised only 7 per cent of the British male labour force as 'professional'. The OPCS
put 19 per cent as 'employers and managers', but the majority of men still occupy manual
work categories (*General Household Survey, 1990*, Table 2.10, HMSO.
4   An important caveat should be made here, though I lack space to develop it fully. While
Bell is no doubt correct to argue that a good deal of service sector work is informational,
we need to question the 'professional' – and even white-collar – nature of much of this.
We can acknowledge a significant growth of professional occupations in the service
sector, but need to remember that this is from a small base point. Even at the upper levels
estimates of professional occupations are no more than 20 per cent of the total workforce.
Not only this, while there may be a long term trend towards more professional jobs,
more recent patterns question its continuation. Katherine Newman, for instance, in
commenting on the 'employment boom in the service sector', observes that:

> What is often missed in this laudatory portrait is the low-wage character of the
> American 'job machine'. Services ranging from fast food to banking, from child care

to nursing home attendants, have burgeoned. In most of these growth areas, however, the wage structure has been unfavourable. A small number of professional jobs that pay well have been swamped by minimum wage positions. About 85 per cent of the new jobs created in the 1980s were in the lowest paying industries – retail trade and personal, business, business, and health services. More than half of the eight million (net) new jobs created in the United States between 1979 and 1984 paid less than $7000 per year (in 1984 dollars). While many of these were part-time jobs (another growth area of dubious value), more than 20 per cent of the year-round, full-time jobs created during this period paid no more than $7000.

(Newman, 1992: 117)

5 INFORMATION AND ADVANCED CAPITALISM: HERBERT SCHILLER

1   By which I mean Marxist-influenced rather than committed to Marxist analysis.

6 INFORMATION MANAGEMENT AND MANIPULATION: JÜRGEN HABERMAS AND THE DECLINE OF THE PUBLIC SPHERE

1   I exclude the 800 or so academic libraries in the UK since, servicing students and researchers, they have significantly different purposes to public libraries. Nonetheless, there are important overlaps (e.g. co-operation between libraries across the sectors, accessibility to academic libraries by members of the public who live in the locality) and a full review would want to consider academic libraries as an integral part of the British library infrastructure.
2   Having been amongst the first to make this criticism, perhaps this is why, now that the Right have taken it aboard, the Left 'seems to have no way of speaking up for the public service ethic, and there is a new breed of know-nothing (or guilt-ridden) councillor, who believes (or affects to believe) that knowledge is by definition "elitist", books "middle class" and English literature "racist"' (Samuel, 1992: 17).
3   The top ten business sponsors for 1991–1992, with projects they supported, were:

1. *Royal Insurance* (£2.1 million to Royal Shakespeare Company).
2. *Digital Equipment* (£2 million for Partners in Arts which included a countrywide National Theatre tour).
3. *British Petroleum* (£1.9 million for the Natural History Museum).
4. *British Telecom* (£1.5 million for Royal Shakespeare touring and Northern Ballet Theatre).
5. *Lloyds Bank* (£1.5 million for Young Musician of the Year).
6. *Barclays Bank* (£1.2 million for Barclays New Stages at the Royal Court theatre).
7. *Natwest Bank* (£1 million for London Contemporary Dance Theatre).
8. *Marks and Spencer* (£600,000 for Scottish International Children's Festival).
9. *Shell UK* (£500,000 for London Symphony Orchestra).
10. *Prudential Corporation* (£380,000 for Great Orchestra series).

(*Guardian*, 25 March, 1992: 38)

4   The Government Statistical Services – the key element of government information – is co-ordinated by the Central Statistical Office and includes the statistics divisions of each major government department, the OPCS (Office of Population, Censuses and Surveys) and the Business Statistics Office.
5   The seriousness of these developments is acknowledged in an Office of Technology

222

Assessment (1988) report on *Federal Information Dissemination in an Electronic Age*. Recognising an escalating take-up of electronic forms of storing and accessing government information, this report identifies the issue of 'public access' as the focal concern.

6 These heavy-handed attempts to staunch leaks embarrassing to government went hand in hand with contrived leaks when these supported government, a point made by E. P. Thompson (1980) in a devastating review of the career of right-wing journalist Chapman Pincher. Here Pincher is depicted as a 'sort of common conduit' for governments to 'have leaked their official secrets, scandals and innuendoes' (p.113) into the 'Fleet Street urinals' (p.127). Such leaking is, attests Thompson, 'an exercise by our superiors in the management of news' (p.115).

7 INFORMATION AND RESTRUCTURING: BEYOND FORDISM?

1 Given that Henry Ford's methods originated in the United States in the second decade of the century this could hardly be so (cf. Meyer, 1981).
2 Sidney Pollard (1983) calculates that between 1950 and 1970 Gross Domestic Product per head rose over 50 per cent, an annual rate well over 2 per cent (p.275).
3 Arthur Marwick (1982) demonstrates that average weekly earnings rose 130 per cent between 1955 and 1969; over the same period retail prices rose only 63 per cent. Moreover, while prices of food and other necessities rose steadily, many consumer goods such as cars, televisions and washing machines actually cost less (p.118).
4 Eric Hobsbawm (1968) observes that in 1957 the British people owed a collective instalment debt of £369 million which grew by 1964 to some £900 million, an expansion of almost 250 per cent (p.225).
5 Late in 1973 OPEC (Organisation of Petroleum Exporting Countries) raised the price of crude well over 200 per cent.
6 The market for dollar deposits and loans outside the USA, and therefore stateless.
7 James Curry (1993) writes: 'Flexible specialisation . . . represents a sort of left liberal Horatio Algerism which has developed out of the socio-economic life experience of its proponents. Many of the policy-makers, particularly the academics, who advocate flexible specialisation are basically producer-entrepreneurs themselves, who rely on loosely formed communities of similar producers as well as associated producers within the university and the university system. The bonds of trust, sources and exchanges of information, communication linkages, and physical agglomeration which flexible specialisation's proponents cite so often have a striking resemblance to many contemporary university communities' (p.118).

8 INFORMATION AND POSTMODERNISM

1 There is no direct translation, but it evokes sensuality and eroticism which the usual translation 'bliss' doesn't quite capture (cf. Barthes, 1976).
2 This is a knowingness that advertisers share with audiences, in these postmodern times, hence the large number of advertisements which are self-mocking, tongue in cheek and ironic, ridiculing the idea that anyone would be persuaded to buy from just watching them.
3 Examples of the hyper-real abound. As I write (May 1994) I read that tourism in Krakow, Poland, is experiencing a boom as entrepreneurs reconstruct memorials to the Holocaust. The impetus here is not historical memory (however unreliable that might be); rather it is a response to the Hollywood movie *Schindler's List*. Sites for reconstruction are being

identified and built on that echo Spielberg's version (for example a hoarding outside a bookshop invites tourists to 'Visit Places in Schindler's List').

9 INFORMATION AND URBAN CHANGE: MANUEL CASTELLS

1   Residues of Marxism have often been discerned in Bell's theories. Victor Ferkiss (1979) considered them 'quasi-Marxist' (p.74), Ian Miles (1978) a form of 'vulgarised Marxism' (p.70), Seymour Martin Lipset (1981) 'an apolitical Marxism' (p.22) and Andrew Schonfield (1969) 'fairly familiar Marxist stuff' (p.20).

2   Family Expenditure Survey (1994) data suggest that between 1979 and 1991 the richest 10 per cent of the population increased their real income by over 60 per cent; conversely the bottom 10 per cent suffered a 17 per cent fall in real income over the same period. These figures were confirmed by Alissa Goodman and Steven Webb (1994) in an Institute of Fiscal Studies report which found that the poorest 5 per cent of the population lost 15 per cent of their income between 1979 and 1991 while the top 20 per cent enjoyed an increase of over 50 per cent. Similar findings were reported in an authoritative Rowntree Foundation publication in early 1995. Katherine Newman (1991) confirms a similar pattern in the United States.

3   The sustained migration of the dispossessed, notably towards global cities, from poor countries provides a continuous supply to replenish the underclass and to ensure that wages remain low. John Berger and Jean Mohr (1975) produced a moving book on the experiences of such groups in Europe. Los Angeles is especially marked by immigration because of the combination of acute deprivation from which people are fleeing in areas like Central America and the relative ease of illegally crossing the Californian border. Rieff (1991) comments that, though the United States has a long history of accepting and absorbing immigrants, in Southern California it is as if history has gone into fast-forward. It fuels severe social and cultural conflict (over seventy languages are spoken in LA schools), something frequently conceived as a *Blade Runner* scenario (LA as a babel of non-Anglo gangs). Castells (1994) comments on the migration of ethnic minorities to European cities over the years, again to perform service jobs at minimal wages. With the integration of Europe and national identities being shaken, his expectation is for 'waves of racism and xenophobia' (p.24) to be directed at these minorities (cf. Enzensberger, 1992).

4   The comparisons with pre-democratic South Africa are compelling. Here is Richard Rayner (1992), from Bradford, Yorkshire, on his new home town: 'Los Angeles was a lot like South Africa. The apartheid wasn't enshrined by law, but by economics and geography, and it was just as powerful' (p.232). He continues to underline the racial commonalities of the two regions, adding that 'in Los Angeles I was afraid of blacks in a way I had never been. I behaved in a way that would have disgusted me in New York or London. I was a racist' (ibid.).

5   It is interesting to note that the ghetto poor has generated a voluminous social science literature, but not so the successes of the global city. An exception is described by John Logan and colleagues (1992) who report the research of Ray Pahl on highly paid financial sector professionals. Pahl found here a '£k culture' in which 'moral worth is judged by the number of noughts on the salary cheque. The corollary is that the poor merit their poverty and are despised for it' (p.147).

BIBLIOGRAPHY

Place of publication is London unless otherwise stated.

Abercrombie, Nicholas, Hill, Stephen and Turner, Bryan S. (1986) *Sovereign Individuals of Capitalism*. Allen and Unwin.

Adam Smith Institute (1986) *Ex Libris*. Adam Smith Institute.

Adam Smith Institute (1993) *What Price Public Service? The Future of the BBC*. Adam Smith Institute.

Adams, Valerie (1986) *The Media and the Falklands Campaign*. Macmillan.

Addison, Paul (1982) *The Road to 1945: British Politics and the Second World War* (1975). Quartet.

Adonis, Andrew (1993) 'Whose Line Is It Anyway?', *Financial Times*, 11 October: 15.

Aglietta, Michel (1979) *A Theory of Capitalist Regulation*. New Left Books.

Allan, Alastair J. (1990) *The Myth of Government Information*. Library Association.

Allred, John R. (1978) 'The Purpose of the Public Library: the Historical View' (1972), reprinted in Totterdell (1978).

Anderson, Perry (1990) 'A Culture in Contraflow – Parts1 and 11', *New Left Review* 180 (March–April): 41–78; (182) July–August: 85–137.

Anderson, Benedict (1991), *Imagined Communities: Reflections on the Origin and Spread of Nationalism*. (1983) Verso, 2nd edition.

Aronowitz, Stanley (1989) *Science as Power: Discourse and Ideology in Modern Society*. Macmillan.

Arrow, Kenneth J. (1979) 'The Economics of Information', in Dertouzos and Moses (1979), ch.14: 306–317.

Atkinson, John (1984) *Flexibility, Uncertainty and Manpower Management*. Brighton: University of Sussex, Institute of Manpower Studies.

Atkinson, John and Meager, Nigel. (1986) *New Forms of Work Organisation*. Brighton: University of Sussex, Institute of Manpower Studies.

Auletta, Ken (1982) *The Underclass*. New York: Random House.

Bagdikian, Ben (1987) *The Media Monopoly*. 2nd edition. Boston: Beacon Press.

Bagguley, Paul, Mark-Lawson, Jane, Shapiro, Dan, Urry, John, Walby, Sylvia and Warden, Alan (1990), *Restructuring: Place, Class and Gender*. Sage.

Bailey, Stephen J. (1989) 'Charging for Public Library Services', *Policy and Politics*, 17 (1): 59–74.

Bannister, Nicholas and Tran, Mark (1993) 'BT Ambition Elbows out Small Man', *Guardian*, 5 June: 40.

Barnaby, Frank (1986) *The Automated Battlefield*. Sidgwick and Jackson.

Barnet, Richard J. and Müller, Ronald E. (1975) *Global Reach: The Power of the Multinational Corporations*. Cape.

225

Barnett, Steven and Curry, Andrew (1994), *The Battle for the BBC: A British Broadcasting Conspiracy?* Aurum Press.

Barnouw, Erik (1978) *The Sponsor: Notes on a Modern Potentate.* New York: Oxford University Press.

Barron, Iann and Curnow, Ray (1979) *The Future with Microelectronics: Forecasting the Effects of Information Technology.* Pinter.

Barthes, Roland (1963) *Sur Racine.* Paris: Seuil.

Barthes, Roland (1964) *Essais critiques.* Paris: Seuil.

Barthes, Roland (1966) *Critique et vérité.* Paris: Seuil.

Barthes, Roland (1967) *Writing Degree Zero* (1953). Translated by Annette Lavers and Colin Smith. Cape.

Barthes, Roland (1976) *The Pleasure of the Text.* Translated by Richard Miller.

Barthes, Roland (1979) *The Eiffel Tower and Other Mythologies.* Translated by Richard Howard. New York: Hill and Wang.

Baudrillard, Jean (1975) *The Mirror of Production.* Translated with an Introduction by Mark Poster. St Louis: Telos Press.

Baudrillard, Jean (1979) *Seduction.* Translated by Brian Singer. Macmillan.

Baudrillard, Jean (1981) *For a Critique of the Political Economy of the Sign.* Translated with an Introduction by Charles Levin. St Louis: Telos Press.

Baudrillard, Jean (1983a) *In the Shadow of the Silent Majorities, or, The End of the Social and Other Essays.* Translated by Paul Foss, John Johnson and Paul Patton. New York: Semiotext(e).

Baudrillard, Jean (1983b) *Simulations.* Translated by Paul Foss, Paul Patton and Philip Beitchman. New York: Semiotext(e).

Baudrillard, Jean (1987) *Forget Foucault.* Translated by Nichole Dufresne. New York: Semiotext(e).

Baudrillard, Jean (1988a) *America* (1986). Translated by Chris Turner. Verso.

Baudrillard, Jean (1988b) *Selected Writings.* Edited, with an Introduction, by Mark Poster. Stanford, CA: Stanford University Press.

Baudrillard, Jean (1991) *La Guerre du golfe n'a pas eu lieu.* Paris: Galilée.

Baudrillard, Jean (1992) *L'illusion de la fin, ou, La Grève des événements.* Paris: Galilée.

Baudrillard, Jean (1993a) *Symbolic Exchange and Death (1976).* Translated by Iain Hamilton Grant. Introduction by Mike Gane. Sage.

Baudrillard, Jean (1993b), *The Transparency of Evil: Essays on Extreme Phenomena* (1990). Translated by James Benedict Verso.

Bauman, Zygmunt (1987) *Legislators and Interpreters: On Modernity, Post-Modernity and Intellectuals.* Cambridge: Polity.

Bauman, Zygmunt (1989) *Modernity and the Holocaust.* Cambridge: Polity.

Beharrell, Peter and Philo, Greg (eds) (1977) *Trade Unions and the Media.* Macmillan.

Bell, Daniel (1962) *The End of Ideology: On the Exhaustion of Political Ideas in the Fifties.* Revised edition. New York: Free Press.

Bell, Daniel (1976a), *The Coming of Post-Industrial Society: A Venture in Social Forecasting* (1973) Harmondsworth: Penguin, Peregrine Books.

Bell, Daniel (1976b), *The Cultural Contradictions of Capitalism.* Heinemann.

Bell, Daniel (1979) 'The Social Framework of the Information Society', in Dertouzous and Moses (1979): 163–211.

Bell, Daniel (1980) *Sociological Journeys, 1960–1980.* Heinemann.

Bell, Daniel (1987) 'The World in 2013', *New Society.* 18 December: 31–37.

Bell, Daniel (1989) 'The Third Technological Revolution and Its Possible Socioeconomic Consequences', *Dissent*, Spring: 164–176.

Bell, Daniel (1990) 'Resolving the Contradictions of Modernity and Modernism', *Society*, 27 (3) March–April: 43–50; 27 (4) May–June: 66–75.

Bellah, Robert N., Madsen, Richard, Sullivan, William M., Swidler, Ann and Tipton, Steven

226

M. (1985) *Habits of the Heart: Individualism and Commitment in American Life.* Berkeley: University of California Press.

Beniger, James R. (1986) *The Control Revolution: Technological and Economic Origins of the Information Society.* Cambridge, MA: Harvard University Press.

Benjamin, Bernard (1988) *Accessibility and Other Problems Relating to Statistics used by Social Scientists.* Swindon: Economic and Social Research Council.

Berger, John and Mohr, Jean (1975) *A Seventh Man: A Book of Images and Words about the Experience of Migrant Workers in Europe.* Harmondsworth: Penguin.

Berman, Marshall (1983) *All That is Solid Melts Into Air: The Experience of Modernity* (1982) Verso.

Bernays, Edward L. (1923) *Crystallizing Public Opinion.* New York: Boni and Liveright.

Bernays, Edward L. (1955) *The Engineering of Consent.* Norman: University of Oklahoma Press.

Bernays, Edward L. (1980) *Public Relations* (1952). Norman: University of Oklahoma Press.

Bernstein, Carl (1992) 'Idiot Culture of the Intellectual Masses', *Guardian*, 3 June: 19.

Bijker, Wiebe E., Hughes, Thomas P. and Pinch, Trevor (eds) (1987) *The Social Construction of Technological Systems: New Directions in the History and Sociology of Technology.* Cambridge, MA.: MIT Press.

Blackwell, Trevor and Seabrook, Jeremy (1985) *A World Still to Win: The Reconstruction of the Post-war Working Class.* Faber and Faber.

Blackwell, Trevor and Seabrook, Jeremy (1988) *The Politics of Hope: Britain at the End of the Twentieth Century.* Faber and Faber.

Blitz, James (1993) 'Nightmare for Governments – Foreign Exchange', *Financial Times*, 24 September.

Block, Fred (1990) *Postindustrial Possibilities: A Critique of Economic Discourse.* Berkeley: University of California Press.

Block, Fred and Hirschhorn, Larry (1979) 'New Productive Forces and the Contradictions of Contemporary Capitalism: A Post-Industrial Perspective', *Theory and Society*, 8 (5): 363–395.

Bolton, Roger (1990) *Death on the Rock and Other Stories.* W. H. Allen/Optomen.

Bonefeld, Werner and Holloway, John (eds) (1991) *Post-Fordism and Social Form: A Marxist Debate on the Post-Fordist State.* Macmillan.

Book Marketing Limited (1992) *Book Facts: An Annual Compendium.* Book Marking Limited.

Boorstin, Daniel J. (1962) *The Image, or What Happened to the American Dream.* Harmondsworth: Penguin.

Boulding, Kenneth E. (1971) 'The Economics of Knowledge and the Knowledge of Economics' in Lamberton (1971), ch. 1: 21–36. First published in 1966 in *American Economic Review*, 56 (2): 1–13.

Boyer, Robert (1990) *The Regulation School: A Critical Introduction.* Translated by Craig Charney. New York: Columbia University Press.

Bracken, Paul (1983) *The Command and Control of Nuclear Forces.* New Haven: Yale University Press.

Bradshaw, Della and Taylor, Paul (1993) 'Putting a Price on Research', *Financial Times*, 23 March.

Braun, Ernest and MacDonald, Stuart (1978) *Revolution in Miniature: The History and Impact of Semiconductor Electronics.* Cambridge University Press.

Braverman, Harry (1974) *Labor and Monopoly Capital: The Degradation of Work in the Twentieth Century.* New York: Monthly Review Press.

Briggs, Asa (1985) *The BBC: The First Fifty Years.* Oxford University Press.

British Telecom (1990) *Competitive Markets in Telecommunications: Serving Customers.* British Telecom.

British Telecom (1993) *Report to Our Shareholders*. British Telecom, September.

Broadcasting Research Unit (1986) *The Public Service Idea in British Broadcasting: Main Principles*. Broadcasting Research Unit.

Brock, Gerald W. (1981) *The Telecommunications Industry: The Dynamics of Market Structure*. Cambridge, MA.: Harvard University Press.

Brown, Phillip and Lauder, Hugh (eds) (1992) *Education for Economic Survival: From Fordism to Post-Fordism?* Routledge.

Browning, Harley L. and Singelmann, Joachim (1978) 'The Transformation of the US Labour Force: The Interaction of Industry and Occupation', *Politics and Society*, 8 (3–4): 481–509.

Buck, Nick, Drennan, Matthew and Newton, Kenneth (1992) 'Dynamics of the Metropolitan Economy', in Fainstein *et al.* (1992), ch.3: 68–104.

Budd, Leslie and Whimster, Sam (eds) (1992) *Global Finance and Urban Living: A Study of Metropolitan Change*. Routledge.

Bulmer, Martin (ed.) (1985) *Essays on the History of British Sociological Research*. Cambridge University Press.

Bulmer, Martin (ed.) (1989) *The Goals of Social Policy*. Unwin Hyman.

Burnham, David (1983) *The Rise of the Computer State*. Weidenfeld and Nicolson.

Burns, Tom (1977) *The BBC: Public Institution and Private World*. Macmillan.

Burrows, William E. (1986) *Deep Black: Space Espionage and National Security*. New York: Random House.

Butcher, David (1983) *Official Publications in Britain*. Clive Bingley.

Calhoun, Craig (1993) 'Postmodernism as Pseudohistory', *Theory, Culture and Society*, 10 (1) February: 75–96.

Callinicos, Alex (1989) *Against Postmodernism*. Cambridge: Polity.

Camagni, Roberto (ed.) (1991) *Innovation Networks: Spatial Perspectives*. Belhaven Press.

Campbell, Duncan and Connor, Steve (1986) *On the Record: Surveillance, Computers and Privacy – The Inside Story*. Michael Joseph.

Cantor, Bill (1989) *Experts in Action: Inside Public Relations*. Edited by Chester Burger. New York: Longman.

Castells, Manuel (1977) *The Urban Question: A Marxist Approach* (1972) Translated by Alan Sheridan. Cambridge, MA.: MIT Press.

Castells, Manuel (ed.) (1985) *High Technology, Space and Society*. Beverly Hills: Sage.

Castells, Manuel (1989) *The Informational City: Information Technology, Economic Restructuring and the Urban-Regional Process*. Oxford: Blackwell.

Castells, Manuel (1994) 'European Cities, the Informational Society, and the Global Economy', *New Left Review*, 204 (March–April): 18–32.

Castells, Manuel and Hall, Peter (1994) *Technopoles of the World: The Making of Twenty-First-Century Industrial Complexes*. Routledge.

'Census of Employment Results, 1991' (1993) *Employment Gazette*, 101 (4) April: 117–126.

Central Statistical Office (1978) *Annual Abstract of Statistics*. No. 114. HMSO.

Central Statistical Office (1983) *Annual Abstract of Statistics*. No. 119. HMSO.

Central Statistical Office (1993) *Annual Abstract of Statistics*. No. 129. HMSO.

Certeau, Michel de (1984) *The Practice of Everyday Life*. Translated by Steven F. Rendall. Berkeley: University of California Press.

Chandler, Alfred D. Jr. (1977) *The Visible Hand: The Managerial Revolution in American Business*. Cambridge, MA.: Harvard University Press.

Clark, Colin (1940) *The Condition of Economic Progress*. Macmillan.

Clarke, Simon (1988) 'Overaccumulation, Class Struggle and the Regulation Approach', *Capital and Class*, 36: 59–92.

Clarke, Simon (1990a) 'The Crisis of Fordism or the Crisis of Social Democracy', *Telos*, 83 (Spring): 71–98.

Clarke, Simon (1990b) 'New Utopias for Old: Fordist Dreams and Post-Fordist Fantasies', *Capital and Class*, 42 (Winter): 131–155.

Coakley, Jerry (1992) 'London as an International Financial Centre', in Budd and Whimster (1992) ch.2: 52–72.

Cockerell, Michael (1989) *Live from Number 10: The Inside Story of Prime Ministers and Television* (1988). Faber and Faber.

Cockerell, Michael, Hennessy, Peter and Walker, David (1984) *Sources Close to the Prime Minister: Inside the Hidden World of the News Manipulators*. Macmillan.

Connor, Steven (1989) *Postmodernist Culture: An Introduction to Theories of the Contemporary*. Oxford: Blackwell.

Cooke, Philip (1988) 'Modernity, Postmodernity and the City', *Theory, Culture and Society*, 5 (2–3) June: 475–492.

Cooke, Philip (1990) *Back to the Future: Modernity, Postmodernity and Locality*. Unwin Hyman.

Corner, John and Harvey, Sylvia (eds) (1991) *Enterprise and Heritage: Crosscurrents of National Culture*. Routledge.

Creighton, Colin and Shaw, Martin (eds) (1985) *The Sociology of War and Peace*. Macmillan.

Crook, Stephen, Pakulski, Jan and Waters, Malcolm (1992) *Postmodernization: Change in Advanced Society*. Sage.

Crowley, David and Mitchell, David (eds) (1994) *Communication Theory Today*. Cambridge: Polity.

Cultural Trends (1989) Issue 4. Policy Studies Institute, December.

Cultural Trends (1990a) Issue 6. Policy Studies Institute, August.

Cultural Trends (1990b) Issue 8. Policy Studies Institute.

Cultural Trends (1992a) Issue 16. Policy Studies Institute.

Cultural Trends (1992b) Issue 15. Policy Studies Institute.

Cultural Trends (1993) Issue 19. Policy Studies Institute, December.

Curran, James (1990) 'The New Revisionism in Mass Communication Research', *European Journal of Communication*, 5 (2–3) June: 135–164.

Curran, James (1991) 'Mass Media and Democracy: A Reappraisal', in Curran and Gurevitch (1991), ch.5: 82–117.

Curran, James and Gurevitch Michael (eds) (1991) *Mass Media and Society*. Edward Arnold.

Curran, James and Seaton, Jean (1988) *Power without Responsibility: The Press and Broadcasting in Britain*. 3rd edition. Routledge.

Curran, James, Gurevitch, Michael and Woollacott, Janet (eds) (1977) *Mass Communication and Society*. Edward Arnold.

Curry, James (1993) 'The Flexibility Fetish', *Capital and Class*, 50 (Summer): 99–126.

Curtis, Liz (1984) *Ireland: The Propaganda War*. Pluto Press.

Dahrendorf, Ralf (1987) 'The Erosion of Citizenship and its Consequences for Us All', *New Statesman*, 12 June: 12–15.

Dahrendorf, Ralf (1992) 'Footnotes to the Discussion', in D. Smith, ch.5: 55–58.

Dandeker, Christopher (1990) *Surveillance, Power and Modernity: Bureaucracy and Discipline from 1700 to the Present Day*. Cambridge: Polity.

Davies, Nick and Black, Ian (1984) 'Subversion and the State', *Guardian*, 17 April: 19.

Davis, Mike (1990) *City of Quartz: Excavating the Future in Los Angeles*. Verso.

Dawes, Len (1978) 'Libraries, Culture and Blacks', in Gerard: 131–137.

De Landa, Manuel (1991) *War in the Age of Intelligent Machines*. New York: Zone Books.

de Vroey, Michel (1984) 'A Regulation Approach Interpretation of the Contemporary Crisis', *Capital and Class*, 23 (Summer): 45–66.

Dennis, Norman and Erdos, George (1993) *Families without Fatherhood*. Foreword by A. H. Halsey. Institute of Economic Affairs, Health and Welfare Unit, Choice in Welfare, no.12.

Dertouzos, Michael L. and Moses, Joel (eds) (1979) *The Computer Age: A Twenty-Year View*. Cambridge, MA.: MIT Press.

Diamond, Edwin and Bates, Stephen (1984) *The Spot: The Rise of Political Advertising on Television*. Cambridge, MA.: MIT Press.

Dicken, Peter (1992) *Global Shift: The Internationalization of Economic Activity*. 2nd edition. Paul Chapman.

Dickson, David (1974) *Alternative Technology and the Politics of Technical Change*. Fontana.

Dickson, David (1984) *The New Politics of Science*. New York: Pantheon.

Dordick, Herbert S. and Wang, Georgette (1993) *The Information Society: A Retrospective View*. Newbury Park, CA: Sage.

Dordick, Herbert S., Bradley, Helen G. and Nanus, Burt (1981) *The Emerging Network Marketplace*. Norwood, NJ: Ablex.

Dosi, Giovanni, Freeman, Christopher, Nelson, Richard, Silverberg, Gerald and Soete, Luc (eds) (1988) *Technical Change and Economic Theory*. Pinter.

Dreier, Peter (1982) 'The Position of the Press in the US Power Structure', *Social Problems*, 29 (3) February: 298–310.

Drucker, Peter F. (1969) *The Age of Discontinuity*. Heinemann.

Drummond, Phillip and Paterson, Richard (eds) (1985) *Television in Transition*. British Film Institute.

Dunford, M. (1990) 'Theories of Regulation', *Environment and Planning D: Society and Space*, 8 (3) September: 297–321.

Eatwell, John, Milgate, Murray and Newman, Peter (eds) (1987) *The New Palgrave: A Dictionary of Economics, Vol.2*. Macmillan.

Elegant, Robert (1981) 'How to Lose a War', *Encounter*, 57 (2) August: 73–90.

Enzensberger, Hans Magnus (1976) *Raids and Reconstructions: Essays in Politics, Crime and Culture*. Pluto Press.

Enzensberger, Hans Magnus (1992) 'The Great Migration', *Granta*, 42 (Winter): 15–54.

Ernst, Dieter (1983) *The Global Race in Microelectronics: Innovation and Corporate Strategies for a Period of Crisis*. Frankfurt: Campus Verlag.

Evans, Christopher (1979) *The Mighty Micro: The Impact of the Computer Revolution*. Gollancz.

Evans, Harold (1983) *Good Times, Bad Times*. Weidenfeld and Nicolson.

Ewen, Stuart (1976) *Captains of Consciousness: Advertising and the Social Roots of the Consumer Culture*. New York: McGraw-Hill.

Ewen, Stuart (1988) *All Consuming Images: The Politics of Style in Contemporary Culture*. New York: Basic Books.

Ewen, Stuart and Ewen, Elizabeth (1982), *Channels of Desire: Mass Images and the Shaping of American Consciousness*. New York: McGraw-Hill.

Fainstein, Susan S. and Harloe, Michael (1992) 'Introduction: London and New York in the Contemporary World', in Fainstein *et al*. (1992) ch.1: 1–28.

Fainstein, Susan F., Gordon, Ian and Harloe, Michael (eds) (1992) *Divided Cities: New York and London in the Contemporary World*. Oxford: Blackwell.

Family Expenditure Survey (1994) *Households below Average Income*. HMSO.

Featherstone, Mike (1991) *Consumer Culture and Postmodernism*. Sage.

Featherstone, Mike (1992) 'Postmodernism and the Aestheticization of Everyday Life', in Lash and Friedman (1992), ch.11: 265–290.

Ferguson, Marjorie (ed.) (1990) *Public Communication: The New Imperatives, Future Directions for Media Research*. Sage.

Ferkiss, Victor (1979) 'Daniel Bell's Concept of Post-Industrial Society: Theory, Myth, and Ideology', *Political Science Reviewer*, 9 (Fall): 61–102.

Field, Frank (1989) *Losing Out: The Emergence of Britain's Underclass*. Oxford: Blackwell.

Fierman, Jaclyn (1994) 'The Contingency Workforce', *Fortune*, 24 January: 20–25.

230

Fiske, John (1987) *Television Culture*. Methuen.

Fiske, John (1991) 'Postmodernism and Television', in Curran and Gurevitch (1991), ch.3: 55–67.

Fitzgerald, Patrick and Leopold, Mark (1987) *Strangers on the Line: The Secret History of Phone Tapping*. Bodley Head.

Flint, Julie (1992) 'The Real Face of War', *Observer*, 3 March: 9.

Ford, Daniel (1985) *The Button: The Nuclear Trigger – Does It Work?* Allen and Unwin.

Forester, Tom (ed.) (1989) *Computers in the Human Context: Information Technology, Productivity and People*. Oxford: Blackwell.

Foucault, Michel (1979) *Discipline and Punish: The Birth of the Prison* (1975). Harmondsworth: Penguin, Peregrine Books.

Foucault, Michel (1980) *Power/Knowledge: Selected Interviews and Other Writings, 1972–1977*. Brighton: Harvester Press.

Fowles, Jib (ed.) (1978) *Handbook of Futures Research*. Westport, CT: Greenwood Press.

Fox, Stephen (1985) *The Mirror Makers: A History of American Advertising and its Creators* (1984). New York: Vintage Books.

Fox, Stephen (1989) 'The Panopticon: From Bentham's Obsession to the Revolution in Management Learning', *Human Relations*, 42 (8): 717–739.

Franklin, Bob (1994) *Packaging Politics: Political Communications in Britain's Media Democracy*. Edward Arnold.

Freeman, Christopher (1974) *The Economics of Innovation*. Harmondsworth: Penguin.

Freeman, Christopher (1987) *Technology Policy and Economic Performance*. Pinter.

Freeman, Christopher and Perez, Carlota (1988) 'Structural Crises of Adjustment, Business Cycles and Investment Behaviour', in Dosi *et al.*, ch.3: 38–66.

Freeman, Christopher, Clark, John and Soete, Luc (1982) *Unemployment and Technical Innovation: A Study of Long Waves and Economic Development*. Pinter.

Friedman, Thomas (1990) *From Beirut to Jerusalem*. Fontana.

Frisby, David (1992) *Simmel and Since: Essays on Georg Simmel's Social Theory*. Routledge.

Fröbel, Folker, Heinrichs, Jürgen, and Kreye, Otto (1980), *The New International Division of Labour: Structural Unemployment in Industrialised Countries and Industrialisation in Developing Countries*. Translated by Pete Burgess. Cambridge University Press.

Fuchs, Victor R. (1968) *The Service Economy*. New York: Columbia University Press.

Fukuyama, Francis (1992) *The End of History and the Last Man*. Hamish Hamilton.

Future of the BBC (1992) A Consultative Document Presented to Parliament by the Secretary of State for National Heritage. HMSO, November.

Galbraith, John Kenneth (1972) *The New Industrial State*. 2nd edition. Harmondsworth: Penguin.

Gamble, Andrew (1988) *The Free Economy and the Strong State: The Politics of Thatcherism*. Macmillan.

Gandy, Oscar H. Jr. (1993) *The Panoptic Sort: A Political Economy of Personal Information*. Boulder, Co: Westview.

Garnham, Nicholas (1990) *Capitalism and Communication: Global Culture and the Economics of Information*. Sage.

Garrahan, Philip and Stewart, Paul (1992) *The Nissan Enigma: Flexibility at Work in the Local Economy*. Mansell.

Gellner, Ernest (1983) *Nations and Nationalism*. Oxford: Blackwell.

Gellner, Ernest (1992) *Postmodernism, Reason and Religion*. Routledge.

Gerard, David (ed.) (1978) *Libraries in Society*. Clive Bingley.

Gershuny, Jonathan I. (1977) 'Post-Industrial Society: The Myth of the Service Economy', *Futures*, 9 (2): 103–114.

Gershuny, Jonathan I. (1978) *After Industrial Society? The Emerging Self-Service Economy*. Macmillan.

Gershuny, Jonathan I. (1983) *Social Innovation and the Division of Labour*. Oxford University Press.

Gershuny, Jonathan I. and Miles, Ian (1983) *The New Service Economy: The Transformation of Employment in Industrial Societies*. Pinter.

Giddens, Anthony (1981) *The Class Structure of the Advanced Societies*. 2nd edition. Hutchinson.

Giddens, Anthony (1984) *The Constitution of Society: Outline of the Theory of Structuration*. Cambridge: Polity.

Giddens, Anthony (1985) *The Nation State and Violence: Volume Two of a Contemporary Critique of Historical Materialism*. Cambridge: Polity.

Giddens, Anthony (1987) *Social Theory and Modern Sociology*. Cambridge: Polity.

Giddens, Anthony (1990) *The Consequences of Modernity*. Cambridge: Polity.

Giddens, Anthony (1991) *Modernity and Self-Identity: Self and Society in the Late Modern Age*. Cambridge: Polity.

Giddens, Anthony (1992) *The Transformation of Intimacy: Sexuality, Love and Eroticism in Modern Societies*. Cambridge: Polity.

Gilbert, Martin (1989) *Second World War*. Weidenfeld and Nicolson.

Gillespie, Andrew E. (1991) 'Advanced Communications Networks, Territorial Integration and Local Development', in Camagni (1991), ch.11: 214–229.

Glasgow, Douglas G. (1980) *The Black Underclass: Poverty, Unemployment, and Entrapment of Ghetto Youth*. New York: Viking.

Glasgow University Media Group (1985) *War and Peace News*. Milton Keynes: Open University Press.

Goddard, John B. (1992) 'New Technology and the Geography of the UK Information Economy, in Robins (1992), ch. 11: 178–201. First published in 1991 as 'Networks of Transactions', *Times Higher Education Supplement*, 22 February: vi.

Goddard, John B. and Gillespie, Andrew E. (1986) 'Advanced Telecommunications and Regional Economic Development', *The Geographical Journal*, 152 (3) November: 383–397.

Golding, Peter (1990) 'Political Communication and Citizenship: The Media and Democracy in an Inegalitarian Social Order', in Ferguson (1990), ch.5: 84–100.

Golding, Peter (1992) 'Communicating Capitalism: Resisting and Restructuring State Ideology – the Case of "Thatcherism"', *Media, Culture and Society*, 14 (4) October: 503–521.

Golding, Peter and Murdock, Graham (1991) 'Culture, Communications, and Political Economy', in Curran and Gurevitch (1991), ch.1: 15–32.

Goldthorpe, John H. (1971) 'Theories of Industrial Society: Reflections on the Rescrudescence of Historicism and the Future of Futurology', *European Journal of Sociology*, 12 (2): 263–288.

Goode, Kenneth (1926) quoted in Shapiro, Stephen R. (1969) 'The Big Sell – Attitudes of Advertising Writers About Their Craft in the 1920s and 1930s'. PhD thesis, University of Wisconsin.

Goodman, Alissa and Webb, Steven (1994) *For Richer, For Poorer*. Institute for Fiscal Studies.

Gordon, David M., Edwards, Richard and Reich, Michael (1982) *Segmented Work, Divided Workers: The Historical Transformation of Labor in the United States*. Cambridge MA.: Harvard University Press.

Gouldner, Alvin W. (1976) *The Dialectic of Ideology and Technology: The Origins, Grammar and Future of Ideology*. Macmillan.

Gouldner, Alvin W. (1979) *The Future of Intellectuals and the Rise of the New Class*. Macmillan. First published in 1978 as 'The New Class Project' in *Theory and Society*, 6 (2) September: 153–203; 6 (3) November: 343–389.

Gouldner, Alvin W. (1980) *The Two Marxisms: Contradictions and Anomalies in the Development of Theory*. Macmillan.

Government Statistical Services (1981) Cmnd 8236. Privy Council Office: HMSO, April.

Gurevitch, Michael, Bennett, Tony, Curran, James and Woollacott, Jane (eds) (1982) *Culture, Society and the Media*. Methuen.

Habermas, Jürgen (1989), *The Structural Transformation of the Public Sphere: An Inquiry into a Category of Bourgeois Society* (1962). Translated by Thomas Burger with the assistance of Frederick Lawrence. Cambridge: Polity.

Hacking, Ian (1990) *The Taming of Chance*. Cambridge University Press.

Hakim, Catherine (1987) 'Trends in the Flexible Workforce', *Employment Gazette*, 95 (11) November: 549–560.

Hall, Peter and Preston, Paschal (1988) *The Carrier Wave: New Information Technology and the Geography of Innovation, 1846–2003*. Unwin Hyman.

Hall, Stuart (1992) '"Our Mongrel Selves": The Raymond Williams Memorial Lecture', *New Statesman and Society*, 'Borderlands' Supplement, 19 June: 6–8.

Hall, Stuart and Jacques, Martin (eds) (1989) *New Times: The Changing Face of Politics in the 1990s*. Lawrence and Wishart.

Hallin, Daniel C. (1986) *The 'Uncensored War': The Media and Vietnam*. New York: Oxford University Press.

Halsey, A. H. (1989) 'Social Polarisation and the Inner City', in Bulmer (1989).

Hamelink, Cees J. (1982) *Finance and Information: A Study of Converging Interests*. Norwood, NJ: Ablex.

Hamilton, Adrian (1986) *The Financial Revolution: The Big Bang Worldwide*. Harmondsworth: Penguin.

Hanson, Dirk (1982) *The New Alchemists: Silicon Valley and the Microelectronics Revolution*. Boston: Little, Brown.

Hargreaves, Ian (1993) *Sharper Vision*. Demos.

Harloe, Michael, Marcuse, Peter, and Smith, Neil (1992) 'Housing for People, Housing for Profits', in Fainstein *et al.* (1992) ch.7: 175–202.

Harris, Ralph (1978) 'Some Issues in Political Economy', in Gerard (1978): 49–57.

Harris, Robert (1983) *Gotcha! The Media, the Government and the Falklands Crisis*. Faber and Faber.

Harris, Robert (1990) *Good and Faithful Servant: The Unauthorized Biography of Bernard Ingham*. Faber and Faber.

Harrison, J. F. C. (1984) *The Common People: A History from the Norman Conquest to the Present*. Flamingo.

Harvey, David (1988) 'Voodoo Cities', *New Statesman and Society*, 30 September: 33–35.

Harvey, David (1989a) *The Urban Experience*. Oxford: Blackwell.

Harvey, David (1989b) *The Condition of Postmodernity: An Enquiry into the Origins of Cultural Change*. Oxford: Blackwell.

Hawthorn, Jeremy (ed.) (1987) *Propaganda, Persuasion and Polemic*. Edward Arnold.

Haywood, Trevor (1989) *The Withering of Public Access*. Library Association.

Hebdige, Dick (1988) *Hiding in the Light: On Images and Things*. Comedia/Routledge.

Henderson, Jeffrey (1989) *The Globalisation of High Technology Production: Society, Space and Semiconductors in the Restructuring of the Modern World*. Routledge.

Henderson, Jeffrey and Castells, Manuel (eds) (1987) *Global Restructuring and Territorial Development*. Sage.

Hepworth, Mark (1989) *Geography of the Information Economy*. Belhaven Press.

Hersch, Seymour (1986) *'The Target is Destroyed': What Really Happened to Flight 007*. Faber and Faber.

Hewison, Robert (1987) *The Heritage Industry: Britain in a Climate of Decline*. Methuen.

Hickethier, Knut and Zielinski, Siegfried (eds) (1991) *Medien/Kultur*. Berlin: Wissenschaftsverlag Volker Spiess.

Hillyard, Paddy and Percy-Smith, Janie (1988) *The Coercive State: The Decline of Democracy in Britain*. Fontana.

Hirsch, Joachim (1991) 'Fordism and Post-Fordism: The Present Social Crisis and its Consequences', in Bonefeld and Holloway, (1991) ch.2: 8–34.

Hirschhorn, Larry (1984) *Beyond Mechanization: Work and Technology in a Postindustrial Age*. Cambridge, MA.: MIT Press.

Hirst, Paul and Zeitlin, Jonathan (eds) (1989) *Reversing Industrial Decline? Industrial Structure and Policy in Britain and her Competitors*. Oxford: Berg.

Hirst, Paul and Zeitlin, Jonathan (1991) 'Flexible Specialisation versus Post-Fordism: Theory, Evidence and Policy Implications', *Economy and Society*, 20 (1) February: 1–56.

Hoag, Paul W. (1985) 'High-Tech Armaments, Space Militarization, and the Third World', in Creighton and Shaw (1985), ch.4: 73–96.

Hobsbawm, Eric J. (1968) *Industry and Empire: An Economic History of Britain since 1750*. Harmondsworth: Penguin.

Hobsbawm, Eric J. and Ranger, Terence (eds) (1983) *The Invention of Tradition*. Cambridge University Press.

Hollingsworth, Mark and Norton-Taylor, Richard (1988) *Blacklist: The Inside Story of Political Vetting*. Hogarth Press.

Holub, Robert C. (1991) *Jürgen Habermas: Critic in the Public Sphere*. Routledge.

Hood, Neil and Young, Stephen (1979) *The Economics of Multinational Enterprise*. Longman.

Horkheimer, Max and Adorno, Theodor W. (1973) *Dialectic of Enlightenment* (1944). Translated by John Cumming. Allen Lane.

Howard, Robert (1985) *Brave New Workplace*. New York: Viking.

Hutton, Will (1994) 'Markets Threaten Life and Soul of the Party', *Guardian*, 4 January: 13.

Ignatieff, Michael (1991) 'Gradgrind Rules in the Public Libraries', *Observer*, 2 June: 19.

ITAP (Information Technology Advisory Panel) (1983) *Making a Business of Information: A Survey of New Opportunities*. HMSO, September.

Jackson, Paul (1992) quoted in *Guardian*, 6 May p. 3.

James, Louis (1973) *Fiction for the Working Man, 1830–1850: A Study of the Literature Produced for the Working Classes in Early Victorian Urban England* (1963). Harmondsworth: Penguin.

Jameson, Fredric (1991) *Postmodernism, Or, The Cultural Logic of Late Capitalism*. Verso.

Janowitz, Morris (1974) 'Review Symposium: *The Coming of Post-Industrial Society*', *American Journal of Sociology*, 80 (1): 230–236.

Janus, Noreene (1984) 'Advertising and the Creation of Global Markets: The Role of the New Communications Technologies', in Mosco and Wasko (1984): 57–70.

Januszczak, Waldemar (1985) 'The Art World Can't Tell Jacob Duck from Donald', *Guardian*, 28 December.

Januszczak, Waldemar (1986) 'No Way to Treat a Thoroughbred', *Guardian*, 15 February: 11.

Jasani, Bhupendra and Lee, Christopher (1984) *Countdown to Space War*. Eyre Methuen.

Jencks, Charles (1984) *The Language of Post-Modern Architecture*. 4th edition. Academy Editions.

Jencks, Christopher (1991) 'Is the American Underclass Growing?', in Jencks and Peterson (1991): 28–100.

Jencks, Christopher and Peterson, Paul E. (eds) (1991) *The Urban Underclass*. Washington, DC: The Brookings Institution.

Johnson, Robert W. (1986) *Shootdown: The Verdict on KAL 007*. Chatto and Windus.

Jones, Gareth Stedman (1984) *Outcast London: A Study in the Relationship between Classes in Victorian Society* (1971). Harmondsworth: Penguin, Peregrine Books, 1984.

Jones, Trevor (ed.) (1980) *Microelectronics and Society*. Milton Keynes: Open University Press.

Jonscher, Charles (1983) 'Information Resources and Economic Productivity', *Information Economics and Policy*, 1: 13–35.

Karunaratne, Neil Dias (1986) 'Issues in Measuring the Information Economy', *Journal of Economic Issues*, 13 (3): 51–68.

Kavanagh, Dennis (1990) *Thatcherism and British Politics: The End of Consensus?* New edition. Oxford University Press.

Keane, John (1991) *The Media and Democracy*. Cambridge: Polity.

Keating, Peter (ed.) (1976) *Into Unknown England, 1866–1913: Selections from the Social Explorers*. Fontana.

Kellner, Douglas (1989a) *Jean Baudrillard: From Marxism to Postmodernism and Beyond*. Cambridge: Polity.

Kellner, Douglas (1989b) *Critical Theory, Marxism and Modernity*. Cambridge: Polity.

Kellner, Douglas (1990) *Television and the Crisis of Democracy*. Boulder, CO: Westview Press.

Kellner, Hans and Berger, Peter L. (eds) (1992) *Hidden Technocrats: The New Class and New Capitalism*. New Brunswick: Transaction.

Kempson, Elaine, Bryson, Alex and Rowlingson, Karen (1994) *Hard Times? How Poor Families Make Ends Meet*. Policy Studies Institute.

Kennedy, Paul (1988) *The Rise and Fall of the Great Powers: Economic Change and Military Conflict from 1500 to 2000*. Unwin Hyman.

King, Anthony D. (1990) *Global Cities: Post-Imperialism and the Internationalization of London*. Routledge.

Kleinberg, Benjamin S. (1973) *American Society in the Postindustrial Age: Technocracy, Power, and the End of Ideology*. Columbus, OH: Merrill.

Knightley, Phillip (1991) 'Here is the Patriotically Censored News', *Index on Censorship*, 20 (4 & 5) April/May: 4–5.

Kolko, Joyce (1988) *Restructuring the World Economy*. New York: Pantheon.

Kroker, Arthur (1992) *The Possessed Individual: Technology and Postmodernity*. Macmillan.

Kroker, Arthur and Cook, David (1986) *The Postmodern Scene: Excremental Culture and Hyper-Aesthetics*. New York: St Martin's Press.

Kroker, Arthur, Kroker, Marilouise and Cook, David (1989) *Panic Encyclopedia: The Definitive Guide to the Postmodern Scene*. Macmillan.

Kumar, Krishan (1977) 'Holding the Middle Ground: the BBC, the Public and the Professional Broadcaster', in Curran *et al.* (1977), pp. 231–248.

Kumar, Krishan (1978) *Prophecy and Progress: The Sociology of Industrial and Post-Industrial Society*. Allen Lane.

Kumar, Krishan (1986) 'Public Service Broadcasting and the Public Interest', in MacCabe and Stewart (1986): 46–61.

Kumar, Krishan (1987) *Utopia and Anti-Utopia in Modern Times*. Oxford: Blackwell.

Kumar, Krishan (1992) 'New Theories of Industrial Society', in Brown and Lauder (1992): 45–75.

Labour Research (1983) 'Multinationals Better at Exporting Jobs than Goods', 72 (4) April: 97–99.

Lamberton, Donald M. (ed.) (1971) *Economics of Information and Knowledge: Selected Readings*. Harmondsworth: Penguin.

Landes, David S. (1969) *The Unbound Prometheus: Technological Change and Industrial Development from 1750 to the Present*. Cambridge University Press.

Lasch, Christopher (1985) *The Minimal Self: Psychic Survival in Troubled Times* (1984). Pan.

Lash, Scott (1990) *Sociology of Postmodernism*. Routledge.

Lash, Scott and Friedman, Jonathan (eds) (1992) *Modernity and Identity*. Oxford: Blackwell.

Lash, Scott and Urry, John (1987) *The End of Organized Capitalism*. Cambridge: Polity.

Lash, Scott and Urry, John (1994) *Economies of Signs and Space*. Sage.

Lasswell, Harold D. (1941) *Democracy through Public Opinion*. Wisconsin: George Banta Publishing Company (the Eleusis of Chi Omega, 43 1, Part 2).

Lasswell, Harold D. (1977) 'The Vocation of Propagandists' (1934), in Lasswell, Harold D., *On Political Sociology*. Chicago: University of Chicago Press.

Laudon, Kenneth C. (1986) *Dossier Society: Value Choices in the Design of National Information Systems*. New York: Columbia University Press.

Lawson, Hilary (1989) Narrator and Director, *Cooking the Books*. Channel 4 TV programme in *Dispatches* series, broadcast 26 April (First shown Autumn 1988).

Leavis, Frank Raymond (1977) *The Great Tradition* (1948). Harmondsworth: Penguin.

Lebergott, Stanley (1993) *Pursuing Happiness: American Consumers in the 20th Century*. Princeton, NJ: Princeton University Press.

Leigh, David (1980) *The Frontiers of Secrecy: Closed Government in Britain*. Junction Books.

Leigh, David and Lashmar, Paul (1985) 'The Blacklist in Room 105', *Observer*, 15 August: 9.

Leiss, William, Kline, Stephen and Jhally, Sut (1986) *Social Communication in Advertising: Persons, Products and Images of Well-Being*. Toronto: Methuen.

Lewis, Dennis A. and Martyn, John (1986) 'An Appraisal of National Information Policy in the United Kingdom', *Aslib Proceedings*, 38 (1) January: 25–34.

Library Association (1983) *Code of Professional Conduct*. Library Association.

Liebowitz, Nathan (1985) *Daniel Bell and the Agony of Modern Liberalism*. Westport, CT: Greenwood Press.

Lipietz, Alain (1986) 'New Tendencies in the International Division of Labour: Regimes of Accumulation and Modes of Regulation', in Scott and Storper (1986), ch.2: 16–40.

Lipietz, Alain (1987) *Mirages and Miracles: The Crises of Global Fordism*. Verso.

Lipietz, Alain (1993) *Towards a New Economic Order: Postfordism, Ecology and Democracy*. Cambridge: Polity.

Lippmann, Walter (1922) *Public Opinion*. Allen and Unwin.

Lipset, Seymour Martin (1981) 'Whatever Happened to the Proletariat? An Historic Mission Unfulfilled', *Encounter*, 56 (6) June: 18–34.

Locksley, Gareth (1984) 'Public May Lose Out Over BT Share Issue', *Computing*, 29 November: 36.

Logan, John, Taylor-Gooby, Peter and Reuter, Monika (1992) 'Poverty and Income Equality', in Fainstein *et al.* (1992), ch.5: 129–150.

Lumek, Roberta (1984) 'Information Technology and Libraries', *Library Management*, 5 (3): 1–60.

Lynd, Robert S. and Hanson, A. C. (1933) 'The People as Consumers', in *President's Research Committee on Social Trends: Recent Social Trends in the United States*. McGraw Hill: 857–911.

Lyon, David (1988) *The Information Society: Issues and Illusions*. Cambridge: Polity.

Lyotard, Jean-François (1984) *The Postmodern Condition: A Report on Knowledge* (1979). Translated by Geoff Bennington and Brian Massumi. Manchester University Press.

Lyotard, Jean-François (1993) *Political Writings*. Translated by Bill Readings and Kevin Paul Geiman. UCL Press.

Maasoumi, Esfandias (1987) 'Information Theory', in Eatwell *et al.* (1987): 846–851.

MacCabe, Colin and Stewart, Olivia (eds) (1986) *The BBC and Public Service Broadcasting*. Manchester: Manchester University Press.

Macey, David (1994) *The Lives of Michel Foucault* (1993). Vintage.

McGarry, Kevin J. (1981) *The Changing Context of Information*. Clive Bingley.

McGregor, Alan and Sproull, Alan (1992) 'Employers and the Flexible Workforce', *Employment Gazette*, 100 (5) May: 225–234.

Machlup, Fritz (1962) *The Production and Distribution of Knowledge in the United States*. Princeton: NJ: Princeton University Press.

Machlup, Fritz (1980) *Knowledge: Its Creation, Distribution, and Economic Significance, Vol. I: Knowledge and Knowledge Production*. Princeton: NJ: Princeton University Press.

Machlup, Fritz (1984) *Knowledge: Its Creation, Distribution, and Economic Significance, Vol. III: The Economics of Information and Human Capital*. Princeton: NJ: Princeton University Press.

Machlup, Fritz and Mansfield, Una (eds) (1983) *The Study of Information*. New York: Wiley.

McKendrick, Neil, Brewer, John and Plumb, J. H. (1982) *The Birth of a Consumer Society: The Commercialization of Eighteenth-Century England*. Hutchinson.

MacKenzie, Donald A. (1990) *Inventing Accuracy: A Historical Sociology of Nuclear Missile Guidance*. Cambridge, MA.: MIT Press.

McPhail, Thomas L. (1987) *Electronic Colonialism: The Future of International Broadcasting and Communication*. 2nd edition. Beverly Hills: Sage.

Madge, Tim (1989) *Beyond the BBC: Broadcasters and the Public in the 1980s*. Macmillan.

Mann, Kirk (1992) *The Making of an English 'Underclass'*. Milton Keynes: Open University Press.

Manwaring-White, Sarah (1983) *The Policing Revolution: Police Technology, Democracy and Liberty in Britain*. Brighton: Harvester.

Marchand, Roland (1985) *Advertising the American Dream: Making Way for Modernity, 1920–1940*. Berkeley: University of California Press.

Marcuse, Peter (1989) '"Dual City": A Muddy Metaphor for a Quartered City', *International Journal of Urban and Regional Research*, 13 (4): 697–708.

Markusen, Ann and Gwiasda, Vicky (1994) 'Multipolarity and the Layering of Urban Functions in World Cities: New York's Struggle to Stay on Top', *International Journal of Urban and Regional Research*, 18 (2) June:167–190.

Marschak, Jacob (1968), 'Richard T. Ely Lecture: Economics of Inquiring, Communicating, Deciding', *American Economic Review*, 58 (2) May: 1–18.

Marshall, T. H. (1973) *Class, Citizenship and Social Development*. Westport, CT: Greenwood Press.

Martin, Bernice (1992) 'Symbolic Knowledge and Market Forces at the Frontiers of Postmodernism: Qualitative Market Researchers (Britain)', in Kellner and Berger(1992): 111–156.

Martin, James (1978) *The Wired Society*. Englewood Cliffs, NJ: Prentice-Hall.

Martin, William J. (1988) 'The Information Society – Idea or Entity?', *Aslib Proceedings*, 40 (11/12) November–December: 303–309.

Marwick, Arthur (1982) *British Society since 1945*. Harmondsworth: Penguin.

Marx, Gary T. (1988) *Undercover: Police Surveillance in America*. Berkeley: University of California Press.

Massey, Doreen (1991) 'Flexible Sexism', *Environment and Planning D: Society and Space*, 9 (1) March: 31–57.

Massiter, Cathy (1985) 'The Spymasters Who Broke Their Own Rules', *Guardian*, 1 March: 13.

Mattelart, Armand (1979) *Multinational Corporations and the Control of Culture: The Ideological Apparatuses of Imperialism*. Brighton: Harvester.

Mattelart, Armand (1991) *Advertising International: The Privatisation of Public Space*. Translated by Michael Chanan. Comedia.

Mattera, Philip (1985) *Off the Books: The Rise of the Underground Economy*. Pluto.

Mayhew, Henry (1971) *The Unknown Mayhew: Selections from the Morning Chronicle, 1849–50*. Edited by E. P. Thompson and Eileen Yeo. Merlin Press.

Melody, William H. (1987) 'Information: An Emerging Dimension of Institutional Analysis', *Journal of Economic Issues*, 21 (3) September: 1313–1339.

Mercer, Derrik, Mungham, Geoff and Williams, Kevin (1987) *The Fog of War: The Media on the Battlefield*. Heinemann.

Meyer, Stephen, III (1981) *The Five Dollar Day: Management and Social Control in the Ford Motor Company, 1908–1921*. Albany: State University of New York Press.

Meyrowitz, Joshua (1985) *No Sense of Place: The Impact of Electronic Media on Social Behavior*. New York: Oxford University Press.

Middlemas, Keith (1979) *Politics in Industrial Society: The Experience of the British System since 1911*. André Deutsch.

Miles, Ian (1978) 'The Ideologies of Futures', in Fowles (1978): 67–97.

Miles, Ian (1991) 'Measuring the Future: Statistics and the Information Age', *Futures*, 23 (9) November: 915–934.

Miles, Ian *et al.* (1990) *Mapping and Measuring the Information Economy*. Boston Spa: British Library Research and Development Department.

Miliband, Ralph (1974) *The State in Capitalist Society* (1969) Quartet.

Miliband, Ralph (1985) 'The New Revisionism in Britain', *New Left Review*, 150 (March–April): 5–26.

Miliband, Ralph and Panitch, Leo (eds) (1992) *Socialist Register 1992*. Merlin Press.

Miliband, Ralph and Saville, John (eds) (1974) *Socialist Register 1973*. Merlin Press.

Ministry of Defence (1983) *The Protection of Military Information: Report of the Study Group on Censorship*. Cmnd 9122. HMSO, December.

Ministry of Defence (1985) *The Protection of Military Information: Government Response to the Report of the Study Group on Censorship*. Cmnd 9499. HMSO, April.

Mollenkopf, John H. and Castells, Manuel (1991) *Dual City: Restructuring New York*. New York: Russell Sage Foundation.

Monk, Peter (1989) *Technological Change in the Information Economy*. Pinter.

Monopolies and Mergers Commission (1986) *British Telecommunications PLC and Mitel Corporation: A Report on the Proposed Merger*. Cmnd 9715. HMSO, January.

Moore, Nick and Steele, Jane (1991) *Information-Intensive Britain: An Analysis of the Policy Issues*. Policy Studies Institute.

Morgan, Kenneth O. (1990) *The People's Peace: British History, 1945–1989*. Oxford University Press.

Morris, Meaghan (1992) 'The Man in the Mirror: David Harvey's "Condition" of Postmodernity', *Theory, Culture and Society*, 9 (1) February: 253–279.

Morrison, David E. and Tumber, Howard (1988) *Journalists at War: The Dynamics of News Reporting during the Falklands Conflict*. Sage.

Mosco, Vincent (1982) *Pushbutton Fantasies: Critical Perspectives on Videotex and Information Technology*. Norwood, NJ: Ablex.

Mosco, Vincent (1989) *The Pay-Per Society: Computers and Communications in the Information Age: Essays in Critical Theory and Public Policy*. Toronto: Garamond Press.

Mosco, Vincent and Wasko, Janet (eds) (1984) *The Critical Communications Review, vol.2: Changing Patterns of Communications Control*. Norwood, NJ: Ablex.

Mosco, Vincent and Wasko, Janet (eds) (1988) *The Political Economy of Information*. Madison: University of Wisconsin Press.

Moser, Sir Claus (1980) 'Statistics and Public Policy', *Journal of Royal Statistical Society A*, 143 Part 1: 1–31.

Mowlana, Hamid, Gerbner, George and Schiller, Herbert I. (eds) (1992) *Triumph of the Image: The Media's War in the Persian Gulf – A Global Perspective*. Boulder, CO: Westview Press.

Muirhead, Bill (1987) 'The Case for Corporate Identity', *Observer*, 25 October.

Mulgan, Geoff J. (1991) *Communication and Control: Networks and the New Economies of Communication*. Cambridge: Polity.

Murdock, Graham (1982) 'Large Corporations and the Control of the Communications Industries', in Gurevitch *et al.* (1982): 118–150.

Murdock, Graham (1990) 'Redrawing the Map of the Communications Industries: Concentration and Ownership in the Era of Privatization', in Ferguson (1990), ch.1: 1–15.

Murdock, Graham and Golding, Peter (1974) 'For a Political Economy of Mass Communications', in Miliband and Saville (1974): 205–234.

Murdock, Graham and Golding, Peter (1977a) 'Capitalism, Communication and Class Relations', in Curran *et al.* (1977): 12–43.

Murdock, Graham and Golding, Peter (1977b) 'Beyond Monopoly – Mass Communications in an Age of Conglomerates', in Beharrell and Philo (1977): 93–117.

Murray, Charles (1984) *Losing Ground: American Social Policy, 1950–1980.* New York: Basic Books.

Murray, Charles (1989) 'Underclass', *Sunday Times Magazine*, 26 November: 26–44.

Murray, Fergus (1987) 'Flexible Specialization in the Third Italy', *Capital and Class*, 33: 84–95.

Murray, Robin (ed.) (1981) *Multinationals beyond the Market: Intra-Firm Trade and the Control of Transfer Pricing.* Brighton: Harvester Press.

Murray, Robin (1985) 'Benetton Britain: The New Economic Order', *Marxism Today*, November: 28–32.

Naisbitt, John (1984) *Megatrends: Ten New Directions Transforming Our Lives.* Futura.

Neil, Andrew (1983) 'The Information Revolution: the New Freedom Will Mean the End of Old-fashioned Capitalism and Socialism', *The Listener*, 23 June: 2–4, 22.

Newby, Howard (1979) *The Deferential Worker: A Study of Farm Workers in East Anglia* (1977). Harmondsworth: Penguin.

Newby, Howard (1985) *Green and Pleasant Land? Social Change in Rural England.* 2nd edition. Wildwood House.

Newby, Howard (1987) *Country Life: A Social History of Rural England.* Weidenfeld and Nicolson.

Newman, Karin (1986) *The Selling of British Telecom.* Holt, Rinehart and Winston.

Newman, Katherine S (1991), 'Uncertain Seas: Cultural Turmoil and the Domestic Economy', in Wolfe (1991): 112–130.

'New Times' (1988) *Marxism Today*, October: 3–33.

Nguyen, Godefroy Dang (1985) 'Telecommunications: A Challenge to the Old Order', in Sharp (1985): 87–133.

Noam, Eli (1992) *Telecommunications in Europe.* New York: Oxford University Press.

Noble, David F. (1977) *America by Design: Science, Technology and the Rise of Corporate Capitalism.* New York: Knopf.

Noble, David F. (1984) *Forces of Production: A Social History of Industrial Automation.* New York: Knopf.

Nordenstreng, Kaarle (1984) *The Mass Media Declaration of UNESCO.* Norwood, NJ: Ablex.

Nordenstreng, Kaarle and Varis, Tapio (1974) *Television Traffic: A One-Way Street?* Paris: UNESCO (Reports and Papers on Mass Communication, no.70).

Norman, Peter (1993) 'Survey of IMF World Economy and Finance', *Financial Times*, 24 September.

Norris, Christopher (1990) *What's Wrong With Postmodernism: Critical Theory and the Ends of Philosophy.* Hemel Hempstead: Harvester Wheatsheaf.

Norris, Christopher (1992) *Uncritical Theory: Postmodernism, Intellectuals and the Gulf War.* Lawrence and Wishart.

Northcott, Jim and Walling, Annette (1989) *The Impact of Microelectronics.* Pinter.

OECD (Organisation for Economic Co-operation and Development) (1981) *Information Activities, Electronics and Telecommunications Activities: Impact on Employment,*

Growth and Trade. ICCP (Information, Computer Communications Policy Series), Vol.1: 139 pages. Paris: OECD.

OECD (1986) *Trends in the Information Economy*. ICCP Series, no.11: 42 pages. Paris: OECD.

OECD (1988) *The Telecommunications Industry: Challenges of Structural Change*. Paris: OECD.

OECD (1991) *Universal Service and Rate Restructuring in Telecommunications*. ICCP Series, no.23. Paris: OECD.

OECD (1993a) *Financial Market Trends*. No.56. Paris: OECD.

OECD (1993b) *Labour Force Statistics, 1971–1991*. Paris: OECD.

Oettinger, Anthony G. (1980) 'Information Resources: Knowledge and Power in the 21st Century', *Science*, 209, 4 July: 191–198.

Office of Arts and Libraries (1988) *Financing our Library Services: Four Subjects for Debate: A Consultative Paper*. Cmnd. 324. HMSO.

Office of Technology Assessment (1988) *Informing the Nation: Federal Information Dissemination in an Electronic Age*. Washington, DC: US Congress, October.

Office of Technology Assessment (1990) *Critical Connections: Communications for the Future*. Washington, DC: US Congress, January.

Ohmae, Kenichi (1993) 'The Rise of the Regional State', *Foreign Affairs*, 72 (2) Spring: 78–87.

Pahl, Raymond E. (1988) 'Some Remarks on Informal Work, Social Polarization and the Social Structure', *International Journal of Urban and Regional Research*, 12 (2) June: 247–266.

Peacock, Professor Alan (1986) (chairman) *Report of the Committee on Financing the BBC*. Cmnd 9824. HMSO.

Penn, Roger (1990) *Class, Power and Technology: Skilled Workers in Britain and America*. Cambridge: Polity.

Perkin, Harold (1990) *The Rise of Professional Society: Britain since 1880*. Routledge; first published 1989.

Phillips, Melanie (1988) 'Hello to a Harsh Age of Cold Economies, Farewell New Society', *Guardian*, 26 February.

Phillips, Melanie (1989) 'Standing Up to be Counted', *Guardian*, 8 December.

Phillips, Melanie (1990) 'Statistics and the Poverty of Integrity', *Guardian*, 27 July.

Phillips, Melanie (1991) 'Private Lies and Public Servants', *Guardian*, 9 January: 21.

Phillips, Melanie (1993) 'The Lost Generation', *Observer*, 17 October: 23.

Phillips, Melanie, Huhne, Christopher and Fairhall, David (1989) 'How Cards are Stacked', *Guardian*, 15 March: 21.

Pianta, Mario (1988) *New Technologies Across the Atlantic: US Leadership or European Autonomy?* Hemel Hempstead: Harvester Wheatsheaf.

Pick, Daniel (1993) *War Machine: The Rationalisation of Slaughter in the Modern Age*. New Haven and London: Yale University Press.

Pilger, John (1991a) 'Video Nasties', *New Statesman and Society*, 25 January: 6–7.

Pilger, John (1991b) 'Information is Power', *New Statesman and Society*, 15 November: 10–11.

Pilger, John (1992) *Distant Voices*. Vintage.

Piore, Michael and Sabel, Charles (1984) *The Second Industrial Divide*. New York: Basic Books

Pollard, Sidney (1983) *The Development of the British Economy, 1914–1980*. 3rd edition. Edward Arnold.

Pollert, Anna (1988) 'Dismantling Flexibility', *Capital and Class*, 34 (Spring): 42–75.

Pollert, Anna (ed.) (1990) *Farewell to Flexibility*. Oxford: Blackwell.

Pope, Daniel (1983) *The Making of Modern Advertising*. New York: Basic Books.

Porat, Marc Uri (1977a) *The Information Economy: Definition and Measurement* (OT

Special Publication 77–12 (1)). Washington, DC: US Department of Commerce, Office of Telecommunications, May. (Contains executive summary and major findings of the study.)

Porat, Marc Uri (1977b) *The Information Economy: Sources and Methods for Measuring the Primary Information Sector (Detailed Industry Reports).* OT Special Publication 77–12 (2). Washington, DC: US Department of Commerce, Office of Telecommunications, May.

Porat, Marc Uri (1978) 'Communication Policy in an Information Society', in Robinson (1978): 3–60.

Poster, Mark (1990) *The Mode of Information: Poststructuralism and Social Context.* Cambridge: Polity.

Poster, Mark (1994) 'The Mode of Information and Postmodernity', in Crowley and Mitchell (1994): 173–192.

Potter, David (1954) *People of Plenty: Economic Abundance and the American Character.* Chicago: University of Chicago Press.

Potter, Dennis (1994) *Seeing the Blossom: Two Interviews and a Lecture.* Faber and Faber.

Preston, William, Herman, Edward S. and Schiller, Herbert I. (1989) *Hope and Folly: The United States and UNESCO, 1945–1985.* Minneapolis: University of Minnesota Press.

Raban, Jonathan (1974) *Soft City.* Fontana.

Raphael, Adam (1989) 'Members Who Lobby in their Own Interest', *Observer*, 16 April.

Raphael, Adam (1990) 'What Price Democracy?', *Observer* (colour supplement), 14 October: 7–47.

Rayner, D. (1981) *Sir Derek Rayner's Report to the Prime Minister.* Central Statistical Office.

Rayner, Richard (1992) 'Los Angeles', *Granta*, 40: 227–254.

Reith, J. C. W. (Lord) (1949) *Into the Wind.* Hodder and Stoughton.

Report by the Minister for the Arts on Library and Information Matters during 1983 (1983) Cmnd 9109. HMSO.

Richelson, Jeffrey T. and Ball, Desmond (1986) *The Ties That Bind: Intelligence Co-operation Between the UK/USA Countries.* Allen and Unwin.

Rieff, David (1991) *Los Angeles: Capital of the Third World.* Phoenix.

Robins, Kevin (1991a) 'Prisoners of the City: Whatever Could a Postmodern City Be?', *New Formations*, 15 (December): 1–22.

Robins, Kevin (1991b) 'Tradition and Translation: National Culture in its Global Context', in Corner and Harvey (1991), ch.1, pp. 21–44, 236–241.

Robins, Kevin (ed.) (1992) *Understanding Information: Business, Technology and Geography.* Belhaven Press.

Robins, Kevin and Webster, Frank (1985) '"The Revolution of the Fixed Wheel": Information, Technology and Social Taylorism', in Drummond and Paterson (1985): 36–63.

Robins, Kevin and Webster, Frank (1986) 'The Media, the Military and Censorship', *Screen*, 27 (2) March–April: 57–63.

Robins, Kevin and Webster, Frank (1989) *The Technical Fix: Education, Computers and Industry.* Macmillan.

Robins, Kevin, Webster, Frank and Pickering, Michael (1987) 'Propaganda, Information and Social Control', in Hawthorn (1987): 1–17.

Robinson, Glen O. (ed.) (1978) *Communications for Tomorrow: Policy Perspectives for the 1980s.* New York: Praeger.

Rosenau, Pauline Marie (1992) *Post-Modernism and the Social Sciences: Insights, Inroads, and Intrusions.* Princeton, NJ: Princeton University Press.

Ross, Andrew (1991) *Strange Weather: Culture, Science and Technology in the Age of Limits.* Verso.

Ross, George (1974) 'The Second Coming of Daniel Bell', in Miliband and Saville (1974): 331–348.

Roszak, Theodore (1986) *The Cult of Information: The Folklore of Computers and the True Art of Thinking*. Cambridge: Lutterworth Press.

Rowntree Foundation (1995) *Inquiry into Income and Wealth*. York: Joseph Rowntree Foundation, February.

Rubin, Michael Rogers and Huber, Mary Taylor (1986) *The Knowledge Industry in the United States, 1960–1980*. New Haven: Yale University Press.

Rubin, Michael Rogers and Taylor, Mary (1981) 'The US Information Sector and GNP: An Input–Output Study', *Information Processing and Management*, 17 (4) June: 163–194.

Rule, James B. (1973) *Private Lives and Public Surveillance*. Allen Lane.

Runciman, W. G. (1990) 'How many Classes are there in Contemporary British Society?', *Sociology*, 24 (3): 377–396.

Rusbridger, Alan (1987) 'Charge of the Write Brigade', *Guardian*, 7 January.

Rustin, Mike (1990) 'The Politics of Post-Fordism: The Trouble with "New Times"', *New Left Review*, 175 (May–June): 54–77.

Sabel, Charles F. (1982) *Work and Politics: The Division of Labour in Industry*. Cambridge University Press.

Said, Edward W. (1991) *The World, the Text, and the Critic* (1984) Vintage.

Samuel, Raphael (1992) 'No Mythic Golden Age', *New Statesman and Society*, 6 March: 16–17.

Sassen, Saskia (1991) *The Global City: New York, London, Tokyo*. Princeton, NJ: Princeton University Press.

Sassen-Koob, Saskia (1987) 'Growth and Informalization at the Core: A Preliminary Report on New York City', in Smith and Feagin (1987), ch. 6: 138–154.

Saunders, Peter (1990) *A Nation of Home Owners*. Unwin Hyman.

Sayer, Andrew (1989) 'Post-Fordism in Question', *International Journal of Urban and Regional Research*, 13 (4): 666–693.

Sayer, Andrew and Walker Richard (1992) *The New Social Economy: Reworking the Division of Labor*. Cambridge, MA: Blackwell.

Scannell, Paddy (1992) 'Public Service Broadcasting and Modern Public Life', in Scannell, et al. (1992): 278–292. First published in 1989 in *Media, Culture and Society*, 11(2): 135–166.

Scannell, Paddy and Cardiff, David (1991) *A Social History of British Broadcasting, Vol.1, 1922–1939: Serving the Nation*. Oxford: Blackwell.

Scannell, Paddy, Schlesinger, Philip and Sparks, Colin (eds) (1992) *Culture and Power: A 'Media, Culture and Society' Reader*. Sage.

Schement, Jorge R. and Lievroux, Leah (eds) (1987) *Competing Visions, Complex Realities: Aspects of the Information Society*. Norwood, NJ: Ablex.

Schiller, Anita R. and Schiller, Herbert I. (1982) 'Who Can Own What America Knows?', *The Nation*, 17 April: 461–463.

Schiller, Anita R. and Schiller, Herbert I. (1986) 'Commercializing Information', *The Nation*, 4 October: 306–309.

Schiller, Anita R. and Schiller, Herbert I. (1988) 'Libraries, Public Access to Information and Commerce', in Mosco and Wasko (1988), ch.8: 146–166.

Schiller, Dan (1982) *Telematics and Government*. Norwood, NJ: Ablex.

Schiller, Herbert I. (1969) *Mass Communications and American Empire*. New York: Augustus M. Kelley.

Schiller, Herbert I. (1973) *The Mind Managers*. Boston: Beacon Press.

Schiller, Herbert I. (1976) *Communication and Cultural Domination*. New York: International Arts and Sciences Press.

Schiller, Herbert I. (1981) *Who Knows: Information in the Age of the Fortune 500*. Norwood, NJ: Ablex.

Schiller, Herbert I. (1983a) 'The Communications Revolution: Who Benefits?', *Media Development*, 4: 18–20.

Schiller, Herbert I. (1983b) 'The World Crisis and the New Information Technologies', *Columbia Journal of World Business*, 18 (1) Spring: 86–90.

Schiller, Herbert I. (1984a) 'New Information Technologies and Old Objectives', *Science and Public* Policy, December: 382–383.

Schiller, Herbert I. (1984b) *Information and the Crisis Economy.* Norwood, NJ: Ablex.

Schiller, Herbert I. (1985a) 'Beneficiaries and Victims of the Information Age: The Systematic Diminution of the Public's Supply of Meaningful Information', *Papers in Comparative Studies*, 4: 185–192.

Schiller, Herbert I. (1985b) 'Breaking the West's Media Monopoly', *The Nation*, 21 September: 248–251.

Schiller, Herbert I. (1985c) 'Information – A Shrinking Resource', *The Nation*, 28 December,1985/4 January, 1986: 708–710.

Schiller, Herbert I. (1987) 'Old Foundations for a New (Information) Age', in Schement and Lievroux (1987), ch.2: 23–31.

Schiller, Herbert I. (1988) 'Information: Important Issue for '88', *The Nation*, 4–11 July: 1, 6.

Schiller, Herbert I. (1989a) *Culture, Inc.: The Corporate Takeover of Public Expression.* New York: Oxford University Press.

Schiller, Herbert I. (1989b) 'Communication in the Age of the Fortune 500: An Interview with Herbert Schiller', *Afterimage*, November.

Schiller, Herbert I. (1990a) 'Democratic Illusions: An Interview with Herbert Schiller', *Multinational Monitor*, 11 (6) June: 19–22.

Schiller, Herbert I. (1990b) 'An Interview with Herbert Schiller', *Comnotes*, Department of Communication, University of California San Diego, 4 (2) Winter: 1–5.

Schiller, Herbert I. (1991a) 'Public Information Goes Corporate', *Library Journal*, 1 October: 42–45.

Schiller, Herbert I. (1991b) 'My Graduate Education [1946–48], Sponsored by the US Military Government of Germany', in Hickethier and Zielinski (1991): 23–29.

Schiller, Herbert I. (1992) 'The Context of Our Work', *Société Française des Sciences de l'Information et de la Communication.* Huitième Congrès National, Lille, 21 May: 1–6.

Schiller, Herbert I. (1993) 'Public Way or Private Road?', *The Nation*, 12 July: 64–66.

Schiller, Herbert I. and Phillips, Joseph (eds) (1970) *Super-State: Readings in the Military-Industrial Complex.* Urbana: University of Illinois Press.

Schiller, Herbert I., Mahoney, Eileen, Alexander, Laurien, Roach, Colleen and Anderson, Robin (1992) *The Ideology of International Communications.* New York: Institute for Media Analysis.

Schlesinger, Philip (1987) *Putting 'Reality' Together: BBC News.* 2nd edition. Methuen.

Schlesinger, Philip (1991) *Media, State and Nation: Political Violence and Collective Identities.* Sage.

Schonfield, Andrew (1969) 'Thinking about the Future', *Encounter*, 32 (2): 15–26.

Schudson, Michael (1984) *Advertising, The Uneasy Persuasion: Its Dubious Impact on American Society.* New York: Basic Books.

Scott, Allen J. and Storper, Michael (eds) (1986) *Production, Work, Territory: The Geographical Anatomy of Industrial Capitalism.* Boston: Allen and Unwin.

Seabrook, Jeremy (1978) *What Went Wrong? Working People and the Ideals of the Labour Movement.* Gollancz.

Seabrook, Jeremy (1982a) *Unemployment.* Quartet.

Seabrook, Jeremy (1982b) *Working-Class Childhood.* Gollancz.

Seabrook, Jeremy (1988a) *The Leisure Society.* Oxford: Blackwell.

Seabrook, Jeremy (1988b) *The Race for Riches: The Human Cost of Wealth.* Basingstoke: Marshall Pickering.

Seabrook, Jeremy (1990) *The Myth of the Market: Promises and Illusions.* Bideford, Devon: Green Books.

Sennett, Richard (1970) *The Uses of Disorder: Personal Identity and City Life*. Allen Lane.

Servan-Schreiber, Jean-Jacques (1968) *The American Challenge*. New York: Atheneum.

Shaiken, Harley (1985) *Work Transformed: Automation and in the Computer Age*. New York: Holt, Rinehart and Winston.

Shannon, Claude and Weaver, Warren (1964) *The Mathematical Theory of Communication*. Urbana: University of Illinois Press; first published 1949.

Sharp, Margaret (ed.) (1985) *Europe and the New Technologies: Six Case Studies in Innovation and Adjustment*. Pinter.

Sharpe, Richard (n. d., c.1990) *The Computer World: Lifting the Lid off the Computer Industry*. TV Choice/Kingston College of Further Education.

Shaw, Roy (1990) 'An Adjunct to the Advertising Business?', *Political Quarterly*, 61 (4) October–November: 375–380.

Shils, Edward (1975) *Center and Periphery: Essays in Macrosociology*. Chicago: University of Chicago Press.

Sinclair, John (1987) *Images Incorporated: Advertising as Industry and Ideology*. Croom Helm.

Singelmann, Joachim (1978a) 'The Sector Transformation of the Labor Force in Seven Industrialized Countries, 1920–1970', *American Journal of Sociology*, 83 (5): 1224–1234.

Singelmann, Joachim (1978b) *From Agriculture to Services: The Transformation of Industrial Employment*. Beverly Hills: Sage.

Sklair, Leslie (1990) *Sociology of the Global System*. Hemel Hempstead: Harvester Wheatsheaf.

Sklar, Martin J. (1988) *The Corporate Reconstruction of American Capitalism, 1890–1916*. Cambridge University Press.

Sloan, Alfred P. (1965) *My Years with General Motors* (1963). Pan.

Smart, Barry (1992) *Modern Conditions, Postmodern Controversies*. Routledge.

Smith, Anthony (1976) *The Shadow in the Cave: A Study of the Relationship between the Broadcaster, his Audience and the State* (1973). Quartet.

Smith, Anthony (1980) *The Geopolitics of Information: How Western Culture Dominates the World*. Faber and Faber.

Smith, Anthony (1986) *The Ethnic Origins of Nations*. Oxford: Blackwell.

Smith, C. (1989) 'Flexible Specialization, Automation and Mass Production', *Work, Employment and Society*, 3 (2): 203–220.

Smith, David J. (ed.) (1992) *Understanding the Underclass*. Policy Studies Institute.

Smith, Michael P. and Feagin, Joe R. (eds) (1987) *The Capitalist City: Global Restructuring and Community Politics*. Oxford: Blackwell.

Smith, Roger B. (1989) 'A CEO's Perspective of his Public Relations Staff', in Cantor (1989): 18–32.

Smythe, Dallas W. (1981) *Dependency Road: Communications, Capitalism, Consciousness, and Canada*. Norwood, NJ: Ablex.

Snoddy, Raymond (1986) 'An Information Revolution', *Financial Times*, 24 March: 13.

Social Trends (1992) 22. HMSO.

Steinfels, Peter (1979) *The Neoconservatives*. New York: Simon and Schuster.

Stephen, Andrew (1992) 'Flames on Streets of LA', *Observer*, 3 May: 17.

Stigler, George J. (1961), 'The Economics of Information', *Journal of Political Economy*, 69 (3) June: 213–225.

Stiglitz, Joseph E. (1985) 'Information and Economic Analysis: A Perspective', *The Economic Journal (Supplement)*, 95: 21–41.

Stonier, Tom (1990) *Information and the Internal Structure of the Universe: An Exploration into Information Physics*. Springer-Verlag.

Stonier, Tom (1983) *The Wealth of Information: A Profile of the Post-Industrial Economy*. Thames Methuen.

Swingewood, Alan (1977) *The Myth of Mass Culture*. Macmillan.

Taylor, A. J. P. (1965) *English History, 1914–1945*. Oxford University Press.

Taylor, F. W. (1947) *Scientific Management*. New York: Harper and Brothers.

Taylor, Mark C. and Saarinen, Esa (1994) *Imagologies: Media Philosophy*. Routledge.

Tedlow, Richard S. (1979) *Keeping the Corporate Image: Public Relations and Business, 1900–1950*. Greenwich, CT: Jai Press.

Terkel, Studs (1977) *Working: People Talk About What They Do All Day and How They Feel About What They Do*. Harmondsworth: Penguin, Peregrine Books.

Thompson, Edward P. (1978) *The Poverty of Theory and Other Essays*. Merlin.

Thompson, Edward P. (1980) *Writing by Candlelight*. Merlin.

Thorn-EMI (1980) *Annual Report*. Thorn-EMI.

Toffler, Alvin (1980) *The Third Wave*. Collins.

Toffler, Alvin (1990) *Powershift: Knowledge, Wealth, and Violence at the Edge of the 21st Century*. New York: Bantam.

Tomlinson, John (1991) *Cultural Imperialism: A Critical Introduction*. Pinter.

Totterdell, Barry (ed.) (1978) *Public Library Purpose: A Reader*. Clive Bingley.

Touraine, Alain (1971) *The Post-Industrial Society: Tomorrow's Social History; Classes, Conflicts and Culture in the Programmed Society*. New York: Wildwood House.

Townsend, Peter (1981) 'By Restricting the Flow of Information, the Government is Restricting the Right to Free and Open Discussion of the Industrial, Economic and Social Conditions of Britain', *Guardian*, 15 July: 15.

Townsend, Peter (1985) 'Surveys of Poverty to Promote Democracy', in Bulmer (1985), ch.13: 228–235.

Traber, Michael and Nordenstreng, Kaarle (eds) (1992) *Few Voices, Many Worlds: Towards a Media Reform Movement*. World Association for Christian Communication.

Tracey, Michael (1978) *The Production of Political Television*. Routledge.

Tracey, Michael (1983) *A Variety of Lives: A Biography of Sir Hugh Greene*. Bodley Head.

Trachtenberg, Alan (1982) *The Incorporation of America: Culture and Society in the Gilded Age*. New York: Hill and Wang.

Tully, Mark (1993) 'Speech to Radio Academy Festival', transcribed in *Guardian*, 14 July: 6.

Tumber, Howard (1993a) '"Selling Scandal": Business and the Media', *Media, Culture and Society*, 15: 345–361.

Tumber, Howard (1993b) 'Taming the Truth', *British Journalism Review*, 4 (1): 37–41.

Tunstall, Jeremy (1977) *The Media are American: Anglo-American Media in the World*. Constable.

Tunstall, Jeremy (1983) *The Media in Britain*. Constable.

Turner, Bryan S. (ed.) (1990) *Theories of Modernity and Postmodernity*. Sage.

Turner, Bryan S. (1992) *Max Weber: From History to Modernity*. Routledge.

Turner, Stansfield (1991) 'Intelligence for a New World Order', *Foreign Affairs*, 70 (4) Fall: 150–166.

Turner, Stuart (1987) *Practical Sponsorship*. Kogan Page.

Twitchell, James B. (1992) *Carnival Culture: The Trashing of Taste in America*. New York: Columbia University Press.

Tyler, Rodney (1987) *Campaign! The Selling of the Prime Minister*. Grafton Books.

Urry, John (1990) *The Tourist Gaze: Leisure and Travel in Contemporary Societies*. Sage.

Useem, Michael (1984) *The Inner Circle: Large Corporations and the Rise of Business Political Activity in the US and UK*. New York: Oxford University Press.

Useem, Michael (1985) 'The Rise of the Political Manager', *Sloan Management Review*, 27 (Fall): 15–26.

Useem, Michael and Karabel, Jerome (1986) 'Pathways to Top Corporate Management', *American Sociological Review*, 51 (April): 184–200.

Usherwood, Bob (1989) *Public Libraries as Public Knowledge*. Library Association.

Varis, Tapio (1985) *International Flow of Television Programmes*. Paris: UNESCO (Reports and Papers on Mass Communication, no.100).

Vattimo, Gianni (1992) *The Transparent Society* (1989). Translated by David Webb. Cambridge: Polity.

Venturi, Robert (1972) *Learning from Las Vegas*. Cambridge, MA.: MIT Press.

Virilio, Paul (1986) *Speed and Politics*. Translated by M. Polizotti. New York: Semiotext(e).

Walsh, Kevin (1992) *The Representation of the Past: Museums and Heritage in the Post-Modern World*. Routledge.

Weber, Max (1976) *The Protestant Ethic and the Spirit of Capitalism* (1930). Translated by Talcott Parsons. Allen and Unwin.

Webster, Frank and Robins, Kevin (1986) *Information Technology: A Luddite Analysis*. Norwood, NJ: Ablex.

Weizenbaum, Joseph (1984) *Computer Power and Human Reason: From Judgement to Calculation* (1976). Harmondsworth: Penguin.

Wernick, Andrew (1991) *Promotional Culture: Advertising, Ideology and Symbolic Expression*. Sage.

West, William J. (1992) *The Strange Rise of Semi-Literate England: The Dissolution of the Libraries*. Duckworth.

Westergaard, John and Resler, Henrietta (1975) *Class in a Capitalist Society: A Study of Contemporary Britain*. Heinemann.

Whitaker, Reg (1992) 'Security and Intelligence in the Post-Cold War World', in Miliband and Panitch (1992): 111–130.

Wilkinson, Barry (1983) *The Shopfloor Politics of New Technology*. Heinemann.

Williams, Karel, Cutler, Tony, Williams, John and Haslam, Colin (1987) 'The End of Mass Production?', *Economy and Society*, 16 (3) August: 405–439.

Williams, Raymond (1961) *The Long Revolution*. Harmondsworth: Penguin.

Williams, Raymond (1980) 'Advertising: The Magic System', in *Problems in Materialism and Culture*. Verso: 170–195.

Williams, William Appleman (1969) *The Roots of Modern American Empire: A Study of the Growth and Shaping of Social Consciousness in a Marketplace Society*. New York: Random House.

Wilson, David L. (1993) 'Scholars Say Financial Barriers Limit Electronic Access to Federal Data', *Chronicle of Higher Education*, 17 February: A15–17.

Wilson, David M. (1989) *The British Museum: Purpose and Politics*. British Museum Publications.

Wilson, Kevin G. (1988) *Technologies of Control: The New Interactive Media for the Home*. Madison: University of Wisconsin Press.

Wilson, William J. (1987) *The Truly Disadvantaged: Inner City Woes and Public Policy*. Chicago: University of Chicago Press.

Wilson, William J. (1991a) 'Public Policy Research and *The Truly Disadvantaged*', in Jencks and Peterson (1991): 460–481.

Wilson, William J. (1991b) 'Studying Inner-City Social Dislocation: the Challenge of Public Agenda Research', *American Sociological Research*, 56: 1–14.

Wolfe, Alan (ed.) (1991) *America at Century's End*. Berkeley: University of California Press.

Woolgar, Steve (1985) 'Why Not a Sociology of Machines? The Case of Sociology and Artificial Intelligence', *Sociology*, 19 (4) November: 557–572.

Woolgar, Steve (1988) *Science: The Very Idea*. Chichester: Ellis Horwood.

Young, Hugo (1989) *One of Us: A Biography of Margaret Thatcher*. Macmillan.

Young, Hugo (1991) 'Nothing but an Illusion of Truth', *Guardian*, 5 February.

Zuboff, Shoshana (1988) *In the Age of the Smart Machine: The Future of Work and Power*. Heinemann.

Zukin, Sharon (1988) *Loft Living: Culture and Capital in Urban Change.* 2nd edition. Hutchinson/Radius.

Zukin, Sharon (1991a) 'The Hollow Center: U.S. Cities in the Global Era', in Wolfe (1991): 245–261.

Zukin, Sharon (1991b) *Landscapes of Power: From Detroit to Disney World.* Berkeley: University of California Press.

Zukin, Sharon (1992) 'The City as a Landscape of Power: London and New York as Global Financial Capitals', in Budd and Whimster (1992): 195–223.

INDEX